SpringerBriefs in Materials

For further volumes:
http://www.springer.com/series/10111

David C. Joy

Helium Ion Microscopy

Principles and Applications

 Springer

David C. Joy
University of Tennessee
Knoxville, TN
USA

ISSN 2192-1091   ISSN 2192-1105 (electronic)
ISBN 978-1-4614-8659-6   ISBN 978-1-4614-8660-2 (eBook)
DOI 10.1007/978-1-4614-8660-2
Springer New York Heidelberg Dordrecht London

Library of Congress Control Number: 2013946312

Printed on acid-free paper

Springer is part of Springer Science+Business Media (www.springer.com)

Acknowledgments

Thanks to Dr. John Notte (Zeiss) for critically reading this manuscript and for sharing his unique expertise in this field; Dr. Brendan Griffin (University of Western Australia); Dr. Lucille Giannuzzi (L. A. Giannuzzi & Associates LLC); Dr. Joe Michael (Sandia National Laboratory); and Dr. Dale Newbury (NIST) for their specialist insights and generous help.

Contents

1   **Introduction to Helium Ion Microscopy**   1

2   **Microscopy with Ions: A Brief History**   5

3   **Operating the Helium Ion Microscope**   9
     3.1   The Ion Source   9
     3.2   Operating the Ion Microscope 11

4   **Ion–Solid Interactions and Image Formation** 17
     4.1   Comparing Electrons and Ions 17
     4.2   Ion-Generated Secondary Electrons (iSE) 19
     4.3   The Interaction Volume 24
     4.4   Are SE and iSE the Same? 25
     4.5   The Imaging Depth of Field 27
     4.6   Topographic Contrast in the HIM 27
     4.7   Ion Channeling Contrast 28
     4.8   Backscattered Ion Imaging 30
     4.9   Scanning Transmission Ion Microscopy 32
     4.10 Detecting Ion Beam Signals 35
     4.11 Practical Issues in HIM with a GFIS 36

5   **Charging and Damage** 39

6   **Microanalysis with HIM** 43
     6.1   Electrons, Ions, and X-rays 43
     6.2   Time of Flight: Secondary Ion Mass Spectrometry 45

7   **Ion-Generated Damage** 49

8   **Working with Other Ion beams** 51

9   **Patterning and Nanofabrication** 55

Conclusion   57

Appendix   59

Bibliography   61

Index   63

Chapter 1
Introduction to Helium Ion Microscopy

The scanning electron microscope (SEM) has become the most widely used high-performance microscope. However because of the fundamental limitations of electron beams the new technology of ion beam microscopy is being developed. Because it supports research in fields as diverse as semiconductor technology, solid state physics, biological materials, and polymers. In every case, it produces high-resolution, eye-catching images which are easy even for nonspecialists to appreciate and interpret. However, the smallest feature that can be resolved by an SEM is limited by the diameter of the beam of electrons brought to a focus at the specimen surface, and after 50 years of continuous development, the SEM is now close to reaching its absolute limit of imaging performance which will occur when the diameter of the focused electron beam can be reduced in size no further. As first recognized by Young (1803), and later by Fresnel (1826), the diameter "d" of the smallest focused spot that can be obtained in any optical system is limited by diffraction to a size given by the expression

$$d = k.\,\lambda/\alpha \qquad (1.1)$$

Here, λ is the wavelength of the incident radiation, which for an SEM would be electrons with a size of the order of 0.01 nm; α is the semi-convergence angle (in radians) of the beam, and k is a parameter, whose magnitude is of the order of unity. The maximum value of α is physically limited to $\pi/2$ radians (90°) so no feature in the image can ever in theory be much smaller than the magnitude of the wavelength λ. In practice, the value of α must generally be chosen much smaller than $\pi/2$ in order to minimize the effects of lens aberrations, so the practical smallest spot size is typically from 10 to 30 times larger than the original, optimistic estimate. For example, when using a 20 kV electron beam such as that employed in present-day SEMs, the smallest beam spot that can be anticipated will be of the order of 0.5–1.0 nm in diameter (Fig. 1.1).

This barrier to the imaging performance of every electron microscope can now be avoided by substituting ions for electrons. Ions are much heavier than electrons, for example, a single helium ion is 7,300× more massive than an electron, and so as shown in Fig. (1.2), their wavelengths at all energies are substantially smaller than those of an electron with the same energy. Ions such as neon (Ne^+) or gallium

D. C. Joy, *Helium Ion Microscopy*, SpringerBriefs in Materials,
DOI: 10.1007/978-1-4614-8660-2_1, © David C. Joy 2013

Fig. 1.1 ORION NS helium
ion microscope

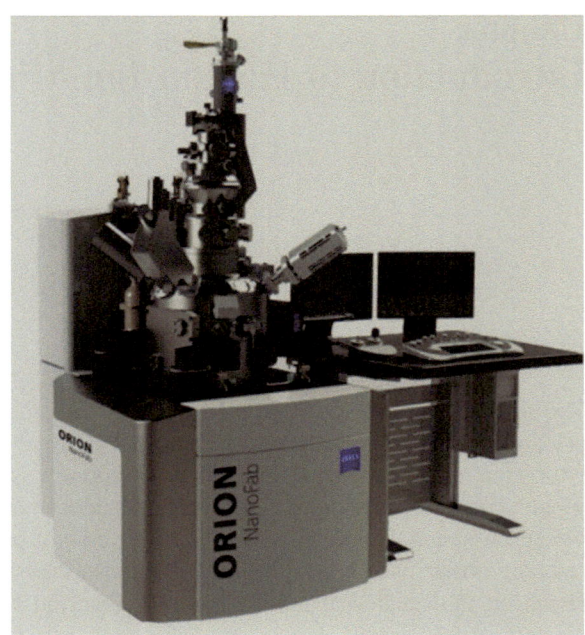

Fig. 1.2 Comparison of the
wavelength of electrons and
selected ions as a function of
their energy

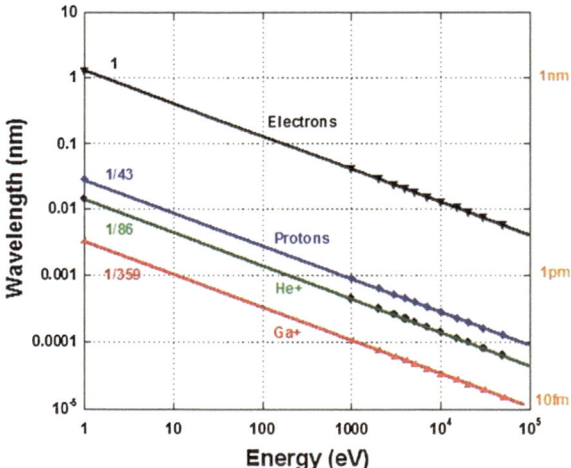

(Ga^+) are heavier by a further factor of ten times or more, and so their wavelength
is correspondingly reduced by another significant factor. As a consequence,
microscopes using ions rather than electrons will no longer have their performance
restricted by the wavelength of the beam, opening up the possibility of ultimately
being able to image objects whose size is subatomic in scale.

Using ions rather than electrons for imaging offers an additional benefit in
which while all electrons are the same, all ions are not. As will be discussed in
more detail later, choosing to use a heavier rather than a lighter ion can have a

profound effect on the microscope performance in areas such as the imaging resolution, the penetration depth of the beam into the specimen at a given energy, the yield of secondary electrons, and the rate at which the ion beam sputters material from the target. The hydrogen ("proton") ion is the nearest in its characteristics to an electron, while more massive ions such as argon (Ar^+) or Ga^+ have the least similarity to electrons. Negative ions can also be used although this is less common. Even larger "pseudoions" such as "buckyballs" which are constructed of sixty carbon atoms could, in principle, be used in an imaging beam, but these special cases have very different characteristics and experimental parameters compared to the more common lighter ions and so will not be discussed here although they do promise to be of considerable research value in the future.

This book firstly provides a brief introduction to the history and development of ion microscopy, then discusses how an ion beam can be generated using a gaseous field ionization source (GFIS) gun and how this performance of this system can be optimized. Next, the optics of the scanning ion microscope will be described, and its operation and imaging performance will be described and evaluated, and the various types of signals available in the ion microscope, their generation and spatial resolution, and the information they provide will be examined. The problem of specimen damage in the ion microscope is then discussed, and the origin of specimen charging and methods for its control in the ion beam are investigated. Finally, the properties and problems associated with using ions other than helium are discussed. In addition, a database of He^+ ion-induced secondary electron yields as a function of the irradiated element or compound, and the beam energy is provided for reference.

Chapter 2
Microscopy with Ions: A Brief History

Every microscope requires a high brightness, reliable, stable source of illumination in order to function, and both the quality and the quantity of the illumination provided will determine, and ultimately limit, the performance of the instrument. Each type of microscope will have its own type of illuminating source. For a high-performance scanning electron or ion microscope, the most desirable property of the beam source is that the source must have a high brightness. The brightness (β) of an emitter is defined as

$$\beta = (\text{signal output})/(\text{area illuminated})/(\text{solid angle subtended by the beam}) \tag{2.1}$$

and has the units of amp/cm^2/steradian. The brightest, and therefore the best, source is the one which emits the highest amount of radiation into the smallest cross-sectional area while still maintaining the most nearly parallel (collimated) beam profile. When the beam is to be used for imaging, it is also desirable that the radiation has the lowest possible energy spread or else each of the individual wavelength components will be brought to a focus at a different distance from the source.

For electron microscopy, the source of choice is a cold field emission gun (CFEG) (Crewe et al. 1968). In this device, a rod of single-crystal tungsten wire is chemically etched to produce a sharp (<100 nm diameter) tip at one end. If the tip is now placed within an ultra-high vacuum container and a voltage is applied between the tip (negative potential) and ground (for the positive potential), then, when the electric field around the tip reaches about 10^7 volts/cm or higher, electrons flow from the tip to ground. The brightness of this beam is of the order of 10^9 amp/cm^2/steradian at 20 keV, which is about 1000× brighter than the signal from a thermionic (hot wire) emitter operated at the same energy. Consequently, the CFEG source is now used in many electron microscopes and has been a major contributor to the dramatic improvements in performance that have been achieved since the 1980s. Used with care, a CFEG in a clean, ultra-high, vacuum system will remain operate for periods of five years or more and require nothing more than

D. C. Joy, *Helium Ion Microscopy*, SpringerBriefs in Materials, DOI: 10.1007/978-1-4614-8660-2_2, © David C. Joy 2013

periodic bake-outs at high temperatures to remove hydrocarbon and other con-
taminants from the end of the tip.

In order to produce a high-performance beam of ions, a related but somewhat
modified approach is employed. This is based on the field ion microscope (FIM), a
device which was developed and perfected by Erwin Müller and his students at
Penn State University in the early 1950s (Muller and Bahadur 1956). Their FIM
was a sealed cylinder containing cryogenically cooled helium (He), chosen
because it is an inert noble gas, causes little sputter damage and modification to
samples, and leaves behind no residue. At one end of the cylinder was a cryogenically cooled metal needle sharpened to an end radius of about 100 nm, and at
the other was a fluorescent viewing screen (Fig. 2.1). When properly set up this
device, operating with a voltage of few kV applied to the tip is capable of pro-
ducing field of the order of 4 volts/angstrom in the vicinity of the tip. In the
presence of such a strong field, any asperities and stray atoms around the tip are
removed by field evaporation, leaving behind a tip region which will now be
approximately spherical in form. Above each of the atoms protruding from the tip,
there is a disk-shaped region within which the field is strong enough to allow
quantum mechanical tunneling of electrons into the needle. Neutral gas atoms
passing through this ionization disk become positively charged and so are accel-
erated away from the needle and can be directed to a suitable imaging screen,
forming an image which displays the crystallographic arrangement of the individual atoms on the needle. Images of this type, initially obtained on October 11,
1955, by Muller's group, were the first direct observation of atomic structures ever
made.

The first ion beam instrument that was also recognizably a "scanning microscope" was developed by Ricardo Levi-Setti, a professor and noted geologist at
the University of Chicago in the 1970s (Levi Setti 1974). His home-built instrument (Fig. 2.2) used a hydrogen (H^+, or Proton) beam operating at about 20 kV
and demonstrated a high level of resolution. The ion source was a modification of
an original field emission electron gun (FEG) developed in Chicago by Professor
Albert Crewe, but also incorporated some ideas from the Muller field ion microscope, in particular the idea of exploiting the very high electric fields that could
induced at an atomically sharp metal tip. The so-called gaseous field ion source
(GFIS) was subsequently adopted by a commercial partnership formed by two
companies, Micrion and JEOL, who were seeking to develop a high-performance

Fig. 2.1 The Muller field ion
microscope—Penn State
University circa 1955

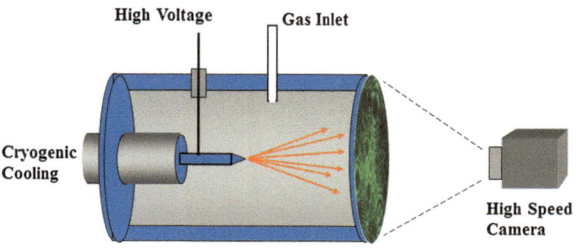

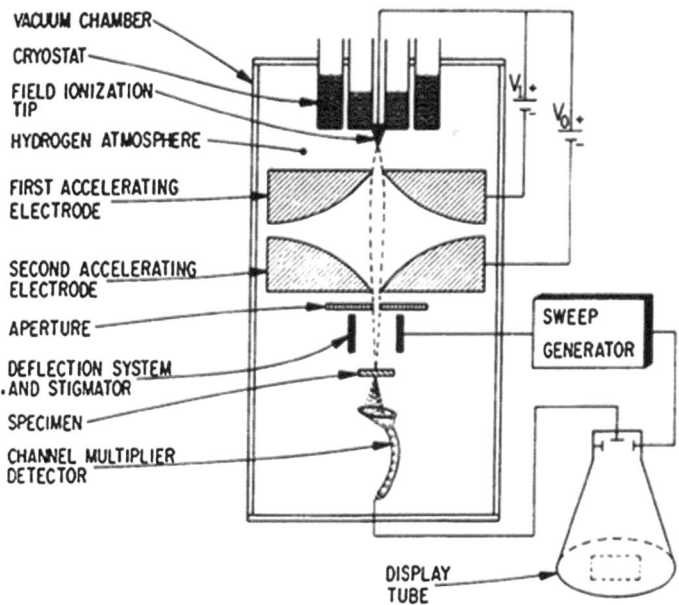

Fig. 2.2 Proton beam scanning microscope designed by Prof. Levi-Setti, University of Chicago, circa 1974

ion source for the semiconductor industry. Subsequently, Micrion was acquired by FEI Inc., before being finally spun off as an independent company in 2002.

In 2005, a new company, the ALIS ("Atomic Level Imaging Systems") Corporation, was incorporated to exploit the technology of the new high-performance helium GFIS source, developed by "Bill" Ward (Ward et al. 2006). This led to the first commercial, high-performance GFIS Helium ion microscope produced by ALIS and bearing the name ORION. Subsequently, the ALIS Company was acquired by Carl Zeiss Inc. in 2007 and is now headquartered in Peabody, MA. Since the first installation, at NIST in Washington DC, about thirty units of the original first-generation ORIONs have been installed, in the USA and across the world. The first example of a second-generation helium ion microscope—the ORION NanoFab (ORION NF)—was delivered to Oak Ridge National Laboratory in November 2012.

The great majority of the references to instrumentation, hardware, and experimental results discussed in this book refer specifically to the Carl Zeiss ORION, or ORION NF HIMs, since these still remain as the only dedicated ion microscopes presently on the market. However, "dual-beam" scanning microscopes, equipped to operate with both electron and ion sources, are now available commercially from several manufacturers and are finding increasing use. However, these instruments most usually employ a Ga^+ (gallium) ion source, chosen because these tools are intended primarily for use in fabrication, and sample preparation for microscopy and microanalysis, and so their spatial resolution can be somewhat limited. A discussion of this type of ion beam instrumentation is outside the scope of this short work, but the volume "Introduction to Focused Ion Beams, Instrumentation, Theory, Techniques, and Practice" (Giannuzzi and Stevie 2005) comprehensively and authoritatively discusses the principles and the application of the dual-beam ion tools for all types of application.

Chapter 3
Operating the Helium Ion Microscope

3.1 The Ion Source

As noted in the previous section, the present ALIS helium ion source is a descendant of the original work based on FIM technology (for a historical overview see Muller and Tsong 1993) although important research in this area has also been carried out by several other prominent groups (e.g. Orloff and Swanson 1977). In order to be suitable for application, in a high-performance particle beam microscope, the source should ideally not only be bright, but also be as compact as possible to ensure mechanical stability, provide highly stable emission over time periods of several hours, be capable of operating at energies at least in the 10–50 keV range, and be capable of being re-formed and then reused multiple times without a significant change in performance. An overview of history of the helium ion microscope can be found in the literature (Economou 2011), while other technical details can be found in the published patents listed at the end of the bibliography.

In the ALIS source, the needle which acts as the emitting electrode is made from a length of proprietary single crystal wire which is about 5 mm long and 0.25 mm in diameter and terminates in a 30° angle cone. In the original HIM, tungsten was the material of choice for the emitter because it is mechanically strong enough not to stretch or break as a result of the high electrostatic fields applied and the associated forces that are generated at the tip. Initially, the end of the tip is covered with a layer of resist as a temporary protection against damage. The tip profile, which at the beginning is roughly hemispherical in form and about 50–100 nm in diameter, is then optimized by processing in the scanning field ion microscope (SFIM) mode—a modern version of Muller's original ion microscope—containing liquid nitrogen at a temperature of about 70 K. While the shaping of the emitter tip is proprietary, there are a number of publications that explore faceting techniques for creating both AFM tips and charged particle sources (Fink 1986; Binh 1988; Kuo et al. 2006).

The tip is now ready to be activated, a process which can be monitored using the SFIM mode. The column is, in effect, set up like the original Muller FIM,

D. C. Joy, *Helium Ion Microscopy*, SpringerBriefs in Materials,
DOI: 10.1007/978-1-4614-8660-2_3, © David C. Joy 2013

projecting the magnified image of the tip region onto a viewing screen. The wire is cooled to near-liquid nitrogen temperatures using helium at a pressure of about 5×10^{-6} torr as the imaging gas. Field evaporation is now performed to remove unwanted atoms, until the SFIM image shows the formation of a clean faceted pyramid structure at the tip. The next step is to carefully optimize the tip shape, so that the electric field is concentrated at the apex of the pyramid while at the same time restricting the field-induced ionization to just the topmost atoms. Starting at about 25 kV, the high voltage applied to the tip region is gradually increased, which results in atoms that are not strongly bound to the emitter being stripped away. This process is allowed to continue until there are just three atoms remaining, arranged in a triangular pattern, at the tip. The "trimer" configuration (Fig. 3.1) maximizes the He^+ ion current because the threefold symmetric arrangement is inherently stable, and because it also guarantees maximum emission in the vertical direction i.e. parallel to the axis of the desired ion beam. A defining aperture—between 5 and 50 microns in diameter in size—is used to select one of the three atoms in the trimer as the emitting source.

The magnitude of the ion current generated varies approximately linearly with helium pressure across a range of about 100:1, with the emission current reaching a maximum value as high as 50–100 pA (equivalent to 0.2–0.5 μA/sterad). The tip temperature is also an important variable, because if the temperature is too low, then the absorption rate of the helium ions is too slow. If, on the other hand, the temperature is too high, then polarized He atoms will have too much kinetic energy to remain bound to the tip long enough to be efficiently ionized. The stability of the beam current is typically better than a few percent over a period of many minutes, although instability can be observed if the vacuum is degraded for

Fig. 3.1 The appearance of an emission trimer as imaged by a SFIM

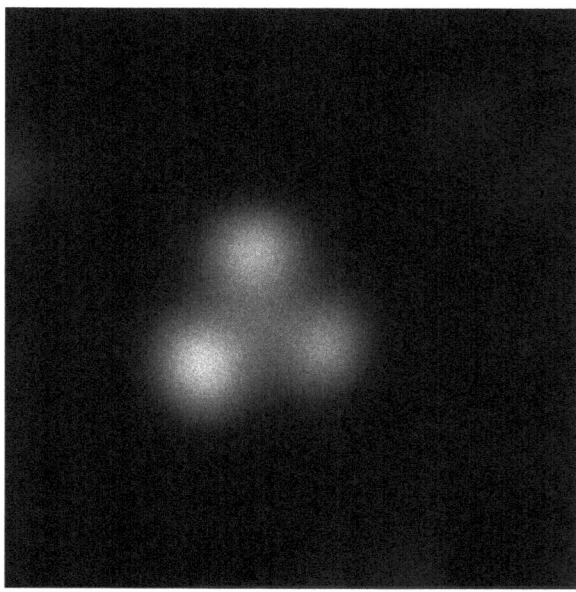

any reason. For extended use, the maximum beam current from a single atom is best limited to about 10 pA, while in routine operation, the emission is more typically set to the order of 1 pA or less. The process which was used to generate the pyramidal tip shape can also be reversed in its effect by increasing the field to about 5 V/Å as at this value, all previous tip atoms are removed. A new pyramid tip can then quickly be rebuilt. In this way, a single source can be periodically reformed and operated for many months. In addition to the manual procedures described above, the ORION HIM provides automated procedures which can either refresh and enhance the performance of an existing trimer, or generate a new trimer, in a few minutes without supervision.

The ALIS GFIS emission source, even before it is considered as the key component of a state-of-the-art microscope, has remarkable properties. The effective source size of the GFIS is of the order of a single atom in diameter, which given the amount of current it provides implies that the source brightness is of the order of 1.10^9 Amps/cm^2/sr at 20 keV—a figure which is comparable with that of a conventional cold field emission source at the same energy. In addition, the beam current available can be varied from a few femtoamps to 10 picoamps or more without the need to change any parameter other than the pressure of the imaging gas. The chromatic energy spread ΔE of the He$^+$ ion beam has been claimed to be less than 1.0 eV (Morgan et al. 2006), a value which is comparable with that of a conventional LaB$_6$ gun (Goldstein et al. 2003) and larger than the energy spread from a cold FEG gun, but substantially smaller than that of a liquid metal ion source. The modest size of this parameter is a significant contribution toward minimizing chromatic aberration in the HIM. The imaging performance of the HIM is also enhanced because the convergence angle "α" of the ion beam is typically of the order of 100 μ-radians, a figure which is 10–100 × smaller than that of any equivalent electron beam tool. Consequently, neither chromatic aberration, which varies as α, nor spherical aberration, which varies as α^3, will degrade the probe size as strongly as they impact on a standard SEM. Unlike conventional electron beam instruments, the HIM does not need, and would not be improved by the use of, an aberration correction system in order to achieve subnanometer resolution performance.

3.2 Operating the Ion Microscope

The variation of the trimer emission current as a function of the voltage applied to the extractor follows the form shown in Fig. 3.2 (Notte et al. 2009). As the voltage increases, the emission current rises, reaches a maximum value, and then falls away before eventually disappearing as the result of field evaporation of the emitter. The voltage at which the emission current reaches its highest value is known as the "best imaging voltage" (BIV), and for Helium, this occurs at a field value of around 4.4 V/A as measured at the apex of the tip. The exact relationship between the voltage and the electric field is determined by the shape of the tip, so sharper tips would achieve the same 4.4 V/A field at a lower voltage, while a blunter tip would require a higher voltage. This field also enhances the performance of the source by acting as a barrier preventing contaminating gas molecules and atoms from reaching the apex of the tip. If the field is allowed to exceed about 5 V/A, it becomes strong enough to remove atoms from the emitter, blunting the tip and partially exposing the next layer of ions. The tip must then be rebuilt for further operation.

As with any other scanning microscope, it is necessary to be able to choose the landing energy of the incident beam onto the specimen. When operating the HIM, this requires that it must be possible to vary the landing energy of the beam voltage independently of the value of the BIV of the emission source. This is achieved with the arrangement shown schematically in Fig. 3.3. Two components are involved—the extractor module which sets up the voltage to achieve the desired BIV condition and the accelerator which determines the landing energy of the ions at the specimen. The extractor unit floats at some energy above ground potential (i.e. zero) determined by the accelerator, which will typically be in the range +25 to +35 kV. The extractor module operates over the range of 0 to −50 keV referenced to the energy of the accelerator and has an output (labeled "ExtV") which is always positive with respect to ground potential.

Figure 3.4 shows the major optical components of an ion microscope, here the Zeiss ORION. The lenses, scanners, and deflectors are all electrostatic in type because ions are only very weakly affected by magnetic fields. The ions generated in the extractor module are then raised to the desired energy by the accelerator and passed through a limiting aperture designed to eliminate any off-axis ions. In normal operation, the beam is then passed through one or two electrostatic lenses. These not only supply some additional demagnification of the probe, if required, but also permit a more precise optical alignment to perform because the lenses can

Fig. 3.2 The variation of trimer ion emission current as a function of the applied voltage. The peak value occurs at "best imaging voltage" (*BIV*) which is the preferred operating condition

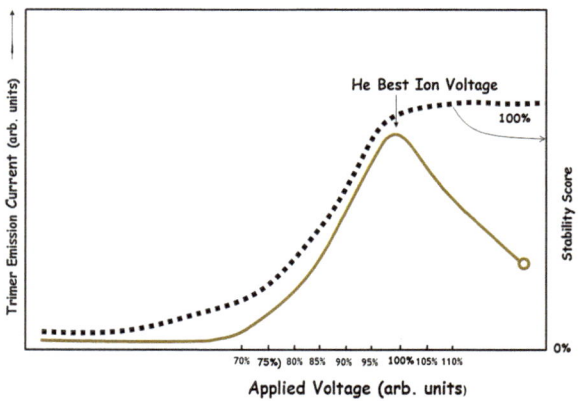

Fig. 3.3 Schematic diagram of the extractor and accelerator components in the HIM

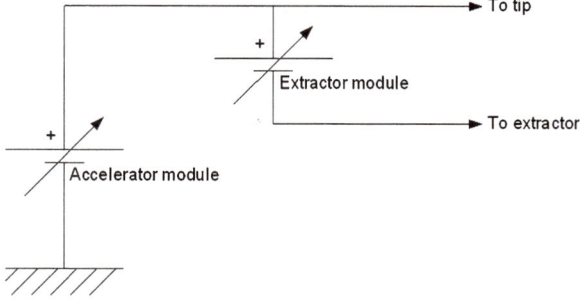

be used to wobble the beam. This is done by superimposing a sinusoidal voltage, with a frequency of about 1 Hz, on top of the static voltage already present on the lenses. The mechanical alignment of the optics can then be optimized by adjusting the X,Y motion controls at the gun so as to bring the beam motion to rest. Similarly, the tilt alignment can be optimized by adjusting the tilt controls so as to eliminate any lateral sweep of the focused beam. Used in combination, these two steps make it possible to achieve the highest performance of which the HIM is capable.

Once all these alignments have been completed, then the operator can quickly, and without the need for any further realignment, vary the incident beam current over a wide range. In addition, by inserting different size apertures into the beam path, the convergence angle can be used to minimize the probe size or to enhance

Fig. 3.4 Simplified cross-sectional diagram of a Zeiss ORION HIM identifying key components

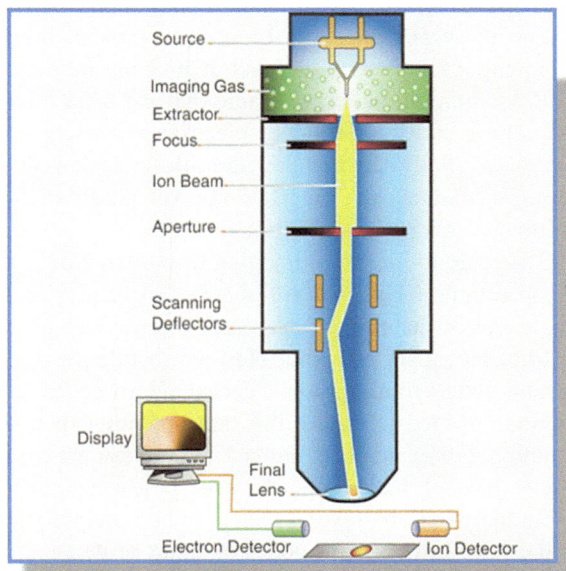

Fig. 3.5 HIM-generated
secondary electron image of
human pancreatic cells. The
beam energy was 36.7 kV,
and the image horizontal field
of view is 800 nm. The
spatial resolution achieved is
about 0.8 nm

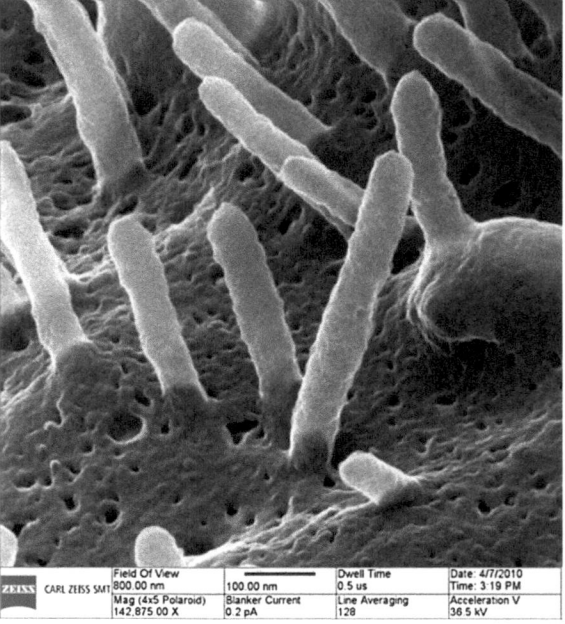

the depth of field of the image. Because the wavelength of the ions is so short, there are only limited diffraction effects, so the smallest diameter apertures in the strip (5 μm) can be used to enhance both the resolution and the depth of field of the image, although at the cost of beam current. Even when using the largest aperture sizes available, the effects of the ion-optical aberrations are still well controlled, and so those apertures can be used to provide higher beam currents while still achieving a relatively high level of imaging performance (Fig. 3.5).

In the long term, it is probable that the helium ion microscope will eventually outperform any conventional electron microscope because the most important parameter of all—the wavelength of the emissions that are used for imaging—is so strongly in its favor, but it is not yet possible to predict what form such an "ultimate" microscope might take.

The route to improved imaging resolution would most simply be accomplished by operating at higher beam energies, because this simultaneously enhances the brightness of the emitter, decreases the wavelength of the ions, and increases the yield of secondary electrons. However, while there are already many conventional "high-voltage transmission electron microscopes" safely in daily use, the precautions needed to ensure the safe operation of a high-voltage HIM look more daunting. For example, a 500-keV helium ion microscope would, in the interests of safety, have to have similarly large physical dimensions to those of a 500-keV TEM and would be several meters in both height and circumference. But because in the HIM only a very modest amount of demagnification of the scanned beam spot is needed to achieve high resolution, the mechanical stability requirements of such an "Ultra-High-Voltage HIM" would have to be stringent if acceptable scanned images, free from artifacts caused by mechanical vibrations, were to be obtained. On the positive side, a HIM would generate much less electro-magnetic radiation than a high-voltage TEM, and it would become possible for ions to generate some limited X-ray emissions for chemical microanalysis.

Chapter 4
Ion–Solid Interactions and Image Formation

Both electron and ion beams can be used to provide a number of different modes of imaging and microanalysis. In every case, and in order to properly optimize and interpret the data generated by the instrument, it is necessary to know something about what kinds of beam interactions are involved, what information may be obtained from each, and how the signal yields and spatial resolution can be optimized in each case. Images whose origins are neither known nor understood can never be any more than just a pretty picture.

4.1 Comparing Electrons and Ions

Energetic ions and electrons can penetrate into, and travel through, solids and liquids. As these particles travel, they lose energy at a rate known as the stopping power, which depends on their energy and on the type of the beam and the density of the medium through which it is traveling. The average maximum depth to which they can descend after entering the specimen is usually designated as being the "beam range." A particle with a short beam range will produce images carrying mostly surface and near subsurface detail, while a beam with high penetration into the sample will be much less affected by what happens at the surface but instead will provide information from interactions that may have occurred at a consider-able depth beneath the surface. Every ion travels along a different, complex, 3-dimensional route to that of every other ion injected into the same material. In a majority of cases, these ions will become embedded somewhere in the specimen, but a finite fraction of the incident ions will always travel back to the surface and then escape, so trying to define a specific "beam range" for ions or electrons, for a given target material and at a specific incident energy, requires a careful definition and, at best, can only represent an averaged value.

The most widely used approximation for estimating beam range is that proposed by Kanaya and Okayama (1972). The beam range R of the ions or electrons is assumed to depend only on the energy E (keV) of the ion and the density ρ of the target material. The Kanaya–Okayama expression then has the form:

D. C. Joy, *Helium Ion Microscopy*, SpringerBriefs in Materials,
DOI: 10.1007/978-1-4614-8660-2_4, © David C. Joy 2013

$$R.\rho = \kappa.E^P \tag{4.1}$$

where R is the ion beam range (nanometers), ρ is the density of the target material (gm/cc), κ is a constant which depends on the particle type (i.e., various ions or electrons), and P is a constant. Note that it is only the density of the medium through which the ions are traveling that determines the particle range, not their atomic number nor their chemical composition, so all materials of the same density will have the same range for a given ion type and energy. Equation (4.3) does not provide any guidance about the lateral spread of the ion beam, but the K–O model is still useful because it provides a reliable and convenient estimate of the scale of the ion–solid interaction.

The difference in beam range between electrons and ions as a function of their energy is immediately evident when the K–O data are plotted (Fig. 4.1). The electron beam range starts at a higher value as compared to ions and then rapidly gets higher still as the accelerating voltage is increased. When using typical SEM energies in the 5–30 keV range, the electron-generated secondary electron (SE) signal consists of information which is derived from the SE2 component, i.e., secondary electrons which have been generated by backscattered electrons as they leave the specimen. Typically, this component is two or three times larger than the yield of SE1 secondary electrons which are generated by the incident beam as it reaches the surface, so the desired detail is only a small component of the total signal and the contrast is poor.

By contrast, the great majority of the ion-generated secondary electron (iSE) signal is generated by incident ions as they first hit the top entrance surface of the sample. Only a very small fraction of the iSE yield comes as a result of back-scattering. The iSE yield is typically 3–5x higher than the corresponding rate for

Fig. 4.1 The beam ranges of electrons and various ions as a function of their energy. Data computed using the Kanaya–Okayama model (Table 4.1)

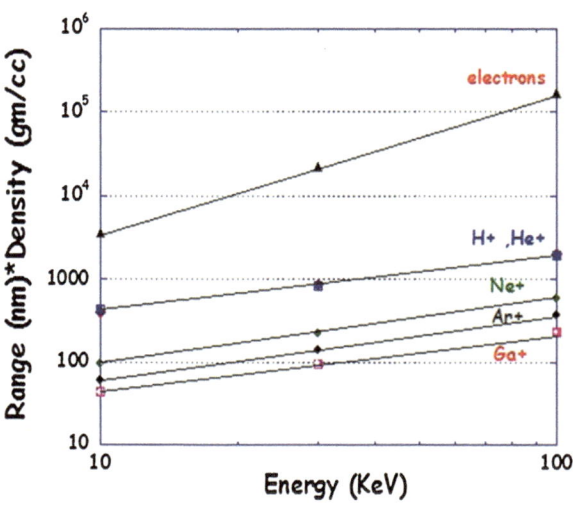

	Particle	k (nm)	P
Table 4.1 Kanaya–Okayama model	Electron	76	1.67
	H^+, He^+	80	0.72
	Ne^+	16	0.72
	Ar^+	10	0.72
	Ga^+	8	0.72

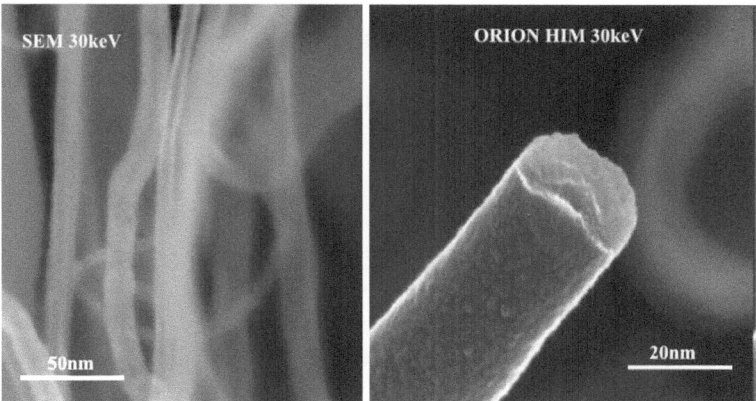

Fig. 4.2 Comparison of SE image of Carbon nanotubes—left-hand side, electron beam generated and, right-hand side, helium ion generated SE

electrons and rises more rapidly as the incident beam energy is increased. iSE images are therefore richer in surface information and also offer superior resolution and a higher signal-to-noise ratio (Fig. 4.2).

4.2 Ion-Generated Secondary Electrons (iSE)

Despite the widespread and highly successful application of secondary electron imaging in SEMs over the past 50 years, this mode of imaging has always tended to be regarded as being most suitable for casual users generating low-resolution images. However, as recent work has now conclusively demonstrated, a strong case can be made for saying that secondary electron imaging is, in fact, the most versatile and universally applicable imaging mode available for charged particle observations. As demonstrated by the recent work of Zhu and his group at

Brookhaven National Laboratory (Zhu et al. 2009), secondary electrons can produce interpretable images of atomic structures displaying subangstrom resolution and contrast detail, which are more than competitive with other, more highly valued, modes of operation. SE imaging is, in fact, the only imaging mode which can seamlessly cover the whole size spectrum from macroscopic to subatomic objects. The combination of ion beams and secondary electrons therefore provides a uniquely versatile and powerful imaging capability.

Secondary electrons (SE) are produced as a result of an ionization event initiated by an energetic incident particle. It is current practice to define secondary electrons as being only those which are emitted from the specimen with energies lower than 50 eV. All the other electrons that may be emitted are considered to have been backscattered (BSE). For the purposes of this discussion, secondary electrons, labeled as being "SE", are those that were generated by electron interactions, while those identified as "iSE" are the product of ion interactions. In general, both electron- and ion-generated secondaries have similar—but not necessarily identical—properties so distinguishing between them on the basis of their origin is helpful.

For most incident ions, and for all targets, the iSE yield (i.e., the average number of iSE generated per incident charged particle) will be higher than that for the electron produced SE yields at any given beam energy. Maximum iSE yields are typically of the order of 5–10 for ions within the 1–10 MeV energy range as compared to a maximum of about 1.5 or less when using electrons. The information carried by the SE and iSE signal components will not be identical even when they come from the same target material because very different operating conditions are required to maximize signal yields and information content in each case. This relationship between the SE and iSE signals can be explained by a simple model.

When an ion or an electron is projected into a material, then as it travels, it loses its energy at a rate, measured in electron volts per angstrom, which is determined by the "stopping power" of the target. The magnitude of the stopping power depends on the energy of the incoming particle and on the physical properties of the medium through which it is traveling. Figure 4.3a shows the variation in stopping power for electrons in chromium, while Fig. 4.3b shows the corresponding stopping power for helium ions in chromium as computed using the SRIM simulation (SRIM 1985). These two stopping power curves can be seen to be quite similar in shape, but while the electron stopping power reaches its maximum magnitude at an energy of several hundred electron volts before falling away, the stopping power of the much more massive He^+ ion only reaches its maximum at a beam energy of about 800 keV. If, however, these same stopping power curves are both plotted as a function of the velocity of the ion or electron, as in Fig. 4.3c, then the two profiles are seen to be very closely similar in behavior, especially when they are normalized to the same peak value.

Bethe (1942) proposed that the rate of production of secondary electrons N_{SE} by an energetic beam of ions or electrons depended directly on the instantaneous magnitude of the stopping power ($-dE/dS$):

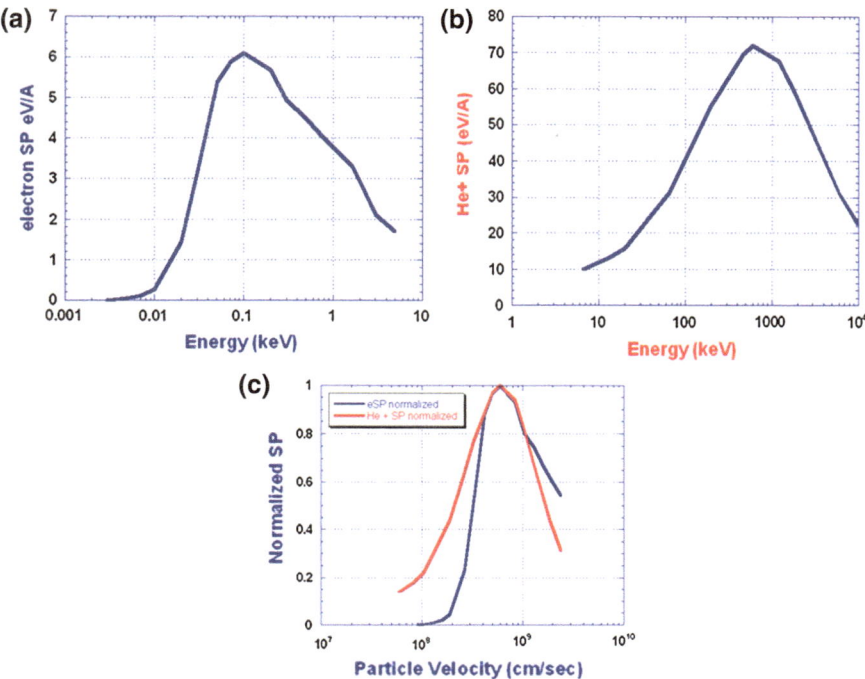

Fig. 4.3 **a** Variation in stopping power for electrons in chromium as a function of energy; **b** SRIME projected stopping power for He$^+$ ions in chrome as a function of energy; **c** combined data from (**a**) and (**b**) but now plotted as a function of particle velocity

$$N(SE) = -1/\varepsilon.\,(dE/dS) \qquad (4.2)$$

where ε is a constant (with units of energy) whose magnitude depends on the choice of the target material, E is the instantaneous energy of the charged particle, and S is the distance traveled by the electron in the material. The SE yield from ions is much higher than that from electrons because the ions deposit their energy more quickly than do electrons. The consequences of changing the incident beam energy are also very different for electron and for ions. The SEM-emitted SE signal is not greatly affected by raising the energy because the increase in beam brightness that is obtained is offset at about the same rate by the reduction in SE production resulting from the fall in stopping power at increasing velocities and by the deeper penetration of the beam into the specimen. Similarly, lowering the SEM beam energy reduces the brightness of the electron gun, but the effect of that loss is offset by the increase in the stopping power, and so again there is little overall change in performance. However, for the ion beam, raising the incident energy increases both the output of ions from the gun and also the stopping power of the target material. The rate of iSE signal generation therefore rises rapidly with energy, greatly enhancing the yield. Reducing the beam energy immediately has the opposite effect, and the iSE signal falls.

A helium ion microscope could, in some respects, be considered as comparable to a very low energy electron microscope. For example, a helium microscope operating at 40 keV is generating ions whose velocity is comparable to that of electrons produced by an SEM operating at only about 5 eV. However, the iSE signal will almost always be larger than the electron SE signal from the same material at the same energy because the stopping power is significantly higher for ions than for electrons. In addition, the ions only travel a short distance in the target and so will on average always be closer to the surface and so have a greater probability of escaping, and there is little or no contribution from backscattered ions most conditions.

Using these ideas, it is now possible to predict how the yield of secondary electrons will vary with energy for both ion and electron generations. The secondary electron generation rate at any depth in the sample is proportional to the instantaneous stopping power at that point and hence is a function of Hi Bipin, HH the velocity of the incoming particle, multiplied by the appropriate yield factor. However, those SE which are generated at depth have to make their way back to the surface of the specimen before they can be detected. The fraction of the SE production ("SE yield") which reaches the surface and so can escape is then

$$\textbf{Yield} = \textbf{0.5} * \textbf{exp}\left(-\textbf{z}/\lambda\right) \qquad (4.3)$$

where λ is the mean free path of the secondary electrons, and so typically is a few nanometers in magnitude, and z is the distance from the point of generation to the nearest point on the sample surface. The production of secondary electrons commences when the electrons or ions first reach the target and continues as long as the electrons or ions travel deeper into the material, stopping only when the incident particle finally comes to rest or escapes into the vacuum once more. Electrons have a longer mean free path and are also more readily backscattered.

Detailed predictions of iSE yields and behavior can be performed using Monte Carlo simulation methods (e.g., Ramachandra et al. 2010; Dapore 2011). For example, Fig. 4.4 shows how the iSE yield generated by a helium beam varies with the energy and with the atomic number of the target. For reference purposes, a listing of helium-generated iSE yields for a variety of elements and materials, as a function of incident ion energy, can be found as an appendix at the end of this book. These data were mostly derived from published experimental data, but because there are often major differences between the yields reported by different observers, even for nominally identical materials and conditions, careful checking of the data is necessary. The tabulated parameters ε, λ also given in the appendix can, when used with the IONiSE Monte Carlo program, reliably predict iSE yields for a wide range of conditions and incident ion beams. However, more, and better, experimental iSE yield data from complex materials are urgently needed.

The magnitude of the iSE yield is almost always higher than the corresponding SE yield even at low ion velocities and becomes significantly larger at velocities

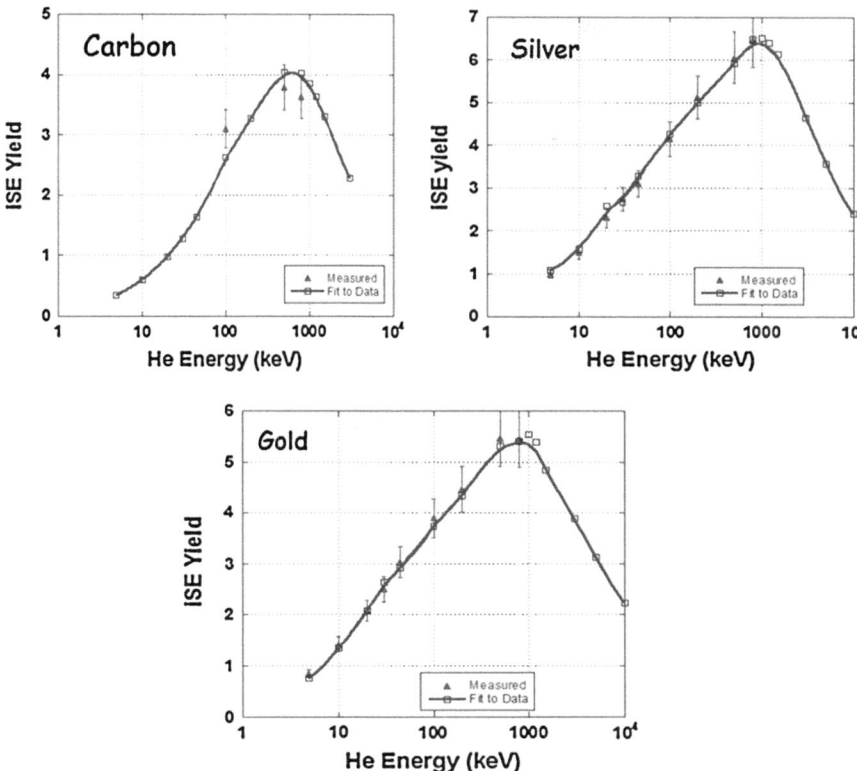

Fig. 4.4 Predicted variation in iSE yield from helium ions as a function of the beam energy and the target material

more typical to those experienced by electrons, i.e., in excess of 10^8 cm/s. One consequence of this is that while electrons generate their SE most efficiently at beam energies of the order of 1 keV, ions generate their iSE most efficiently at incident energies ranging from 50 keV for a He^+ beam, to megavolt levels and beyond for ions such as Ga^+. Because the range of penetration of ions into a solid is usually much less than that for electrons, high beam energy iSE images will offer improved resolution as well as the much superior signal yields as discussed later. Unfortunately, the data available are incomplete and unreliable for a variety of reasons. For example, there is little information on iSE yields generated at beam energies in excess of 50 kV because of the problems associated with generating a stable He^+ (or other ion) beam at such energies, or higher, and the great majority of data so far published deal only with pure elements rather than important compounds of wide interest.

4.3 The Interaction Volume

When the ion beam enters the target material, it begins to scatter, initially about the original forward direction but later into a wider range of angles, to form an interaction volume. The size and shape of this region is important because of its effect on the subsequent behavior of the ions. If the interaction volume is small in size, then the volume within which secondary electrons are produced is tightly localized and favorable for high spatial resolution imaging. On the other hand, a large, diffuse interaction volume will degrade the resolution and contrast SE image. The factors that influence this will include the type of ion being used, its energy, and the nature of the target. Figure 4.5 compares the size and shape of the interaction volume at an incident beam energy of 40 kV and for five different ions (H^+, He^+, Ne^+, Ar^+, and Ga^+) irradiating a molybdenum target. The lightest ion (H^+) has an interaction volume which is almost entirely forward looking and extends downward for about 500 nm from the entrance surface. The heaviest ion (Ga^+) has a maximum penetration of only about 80 nm and is no longer predominantly directed downward but shows significant backscattering to be occurring. For electrons, the scale of the interaction volume increases with beam energy E as about $E^{1.6}$, but for ions, the corresponding increase is only proportional to $E^{0.6}$ and so always remains considerably more confined.

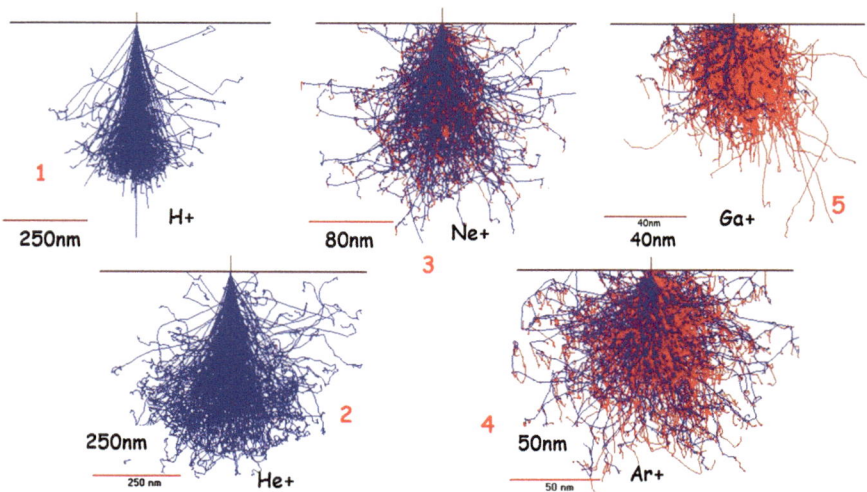

Fig. 4.5 The comparative shape and size of the interaction volumes for H^+, He^+, Ne^+, Ar^+, and Ga^+ in molybdenum at an energy of 40 kV

4.4 Are SE and iSE the Same?

The secondary electrons generated by both incident ions (iSE) or electrons (SE) have energies which lie within the range between zero and 50 eV and which generate images which look similar in information content. However, iSE and SE signals that result are not identical because of important physical differences in how the incident ions and electrons interact with the specimen.

When an electron impinges on the surface of the target material, generation of type SE1 secondaries begins and continues until these SE can no longer find their way back to the entrance surface. The incident electrons will, however, continue to travel into the specimen, but no further secondary electrons emission will be generated unless the trajectory of the electron once again takes it back toward the surface. This will occur if the electron is backscattered. In this case, then, as the electron passes once again through the top surface region, it will generate an additional secondary electron contribution usually designated as "SE2." The yield of these SE2 increases as the energy of the backscattered electron falls, and in addition, as shown in the figure, the backscattered electrons may be moving more parallel to the surface rather than to the vertical. Consequently, in most materials and at most incident beam energies, the SE2 yield is significantly higher than the SE1 yield. These SE2 electrons are not "high resolution" because they have traversed a lengthy arbitrary path through the specimen and so probably carry little or no useful image information. The total SE yield is then

$$SE = SE_1 + SE_2 = SE_1 (1 + \delta\beta) \tag{4.4}$$

where δ is the backscattering coefficient of the specimen and β is a parameter—typically with a value between 3 and 4—which accounts for the enhancement of SE production which occurs as the energy of the incident electrons falls. For most materials, especially those with a high atomic number and for beam energies above a few keV, the SE signal is dominated by the SE_2 component and carries little or no surface information,

In case of an ion beam secondary electron (iSE_1), production again occurs as ions impinge on the sample surface. But because these ions are massive compared to electrons, they are traveling at typically 1 % or less of the velocity of electrons of the same energy. As a result, there is little or no ion backscattering, and consequently, ion-generated iSE2 generation is usually not significant. A comparison of experimentally measured SE with iSE spectra from silicon (Fig. 4.6) shows the nature of the difference very clearly. The electron-generated SE signal spans the energy range from 1 eV up to 50 eV. The corresponding iSE spectrum is initially identical in form, when scaled to match the SE data, from 1 eV up to about 15 eV but then its magnitude fall rapidly with energy and it typically disappears at energy less than 20 eV. The big difference between the iSE and SE spectra is therefore the absence of the backscattered contribution. As a result, the

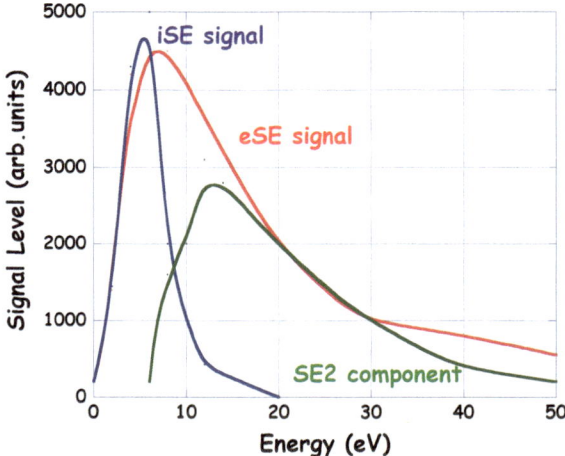

Fig. 4.6 Comparison of SE with iSE energy spectra in the range below 50 eV for silicon

iSE image can be expected to give higher resolution images than those generated by electrons because there will be only a minimal iSE2 contribution reducing image contrast. Ion-beam-generated secondary electron (iSE) images therefore offer better resolution as well as much enhanced surface detail (Fig. 4.7).

Fig. 4.7 iSE image (*left*) of a self-assembled monolayer film of nitro-biphenyl-thiol on a gold substrate, field of view 1 μm, and (*right*) gold evaporated on to carbon, field of view 200 nm. Beam energy is 35 kV. *Note* the high visibility of even monolayer films

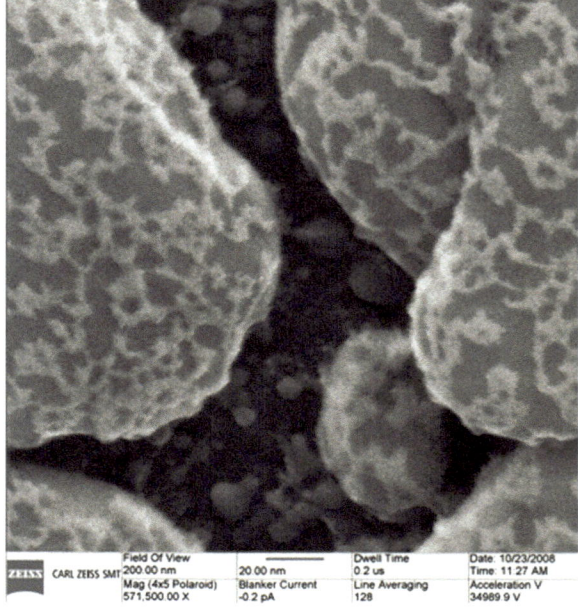

4.5 The Imaging Depth of Field

The experimental advantages of He$^+$ ion-generated secondary electron images are further enhanced by the optics of the HIM itself. The depth of field of an SEM, defined as the distance from the front to the back of a sample within which everything is in focus, is strictly limited because the electron beam has to be highly convergent in order to obtain the best compromise between the diffraction, spherical, and chromatic aberrations of the instrument (Goldstein et al. 2003). Under typical conditions, this requires a convergence angle α of about 10 milliradians or greater, which restricts the magnitude of the depth of the field to just a small fraction of the horizontal field of view of the image.

When using helium, and other ions, the situation is much different because the semi-angle of convergence of the focused beam can now be 50–100x less than in the electron case yet without suffering the effects of diffraction because of the much smaller wavelengths represented by the ion. Reducing the aperture size then directly enhances the depth of field, but using a very small convergence angle will also reduce the available beam current, so a sensible compromise must be determined. Nevertheless, under realistic operating conditions, the imaging depth of field can become comparable in magnitude to the horizontal field of view of the image, so providing a significant improvement in information content from each image.

4.6 Topographic Contrast in the HIM

Topographic contrast is the most versatile and widely used mode of imaging in both the HIM and SEM. "Topography" is the signal contribution that is generated when an incident ion, or electron, beam is scanned across a specimen surface which is either not flat, or is not positioned normal to the beam. The yield of both iSE and SE signals increases with the angle of incidence between the beam and the specimen surface resulting in images which most observers find to be easy to appreciate and interpret because they closely match what we would be observe with our own eyes. The effective source of "light" in this case is the location occupied by the secondary electron detector, while the observer is positioned directly on top of the microscope looking straight downward on to the specimen. This mode of observation accounts for 90 % or more of all the images generated by SEMs because it is effective, is intuitive, and behaves in the same manner across the size scale ranging from millimeters to nanometers. As an additional benefit, and when proper precautions are taken, topographic signal profiles can be interpreted quantitatively for applications such as making three-dimensional models of a surface, or measuring the width or height or spacing, of micro- and nanoscale features in semiconductor devices or nanoscale structures (Joy 2011).

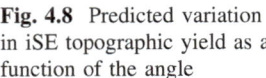

Fig. 4.8 Predicted variation in iSE topographic yield as a function of the angle

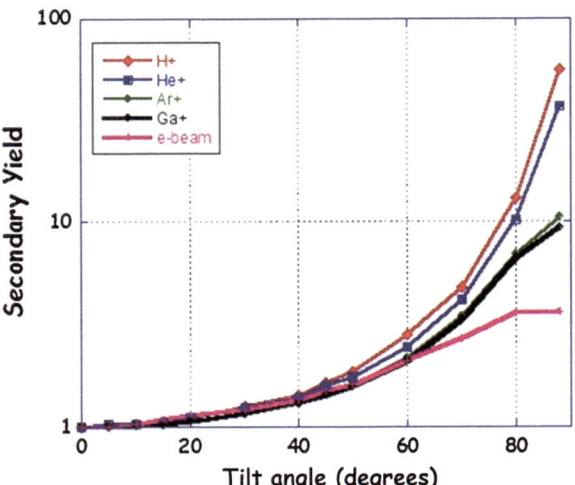

The variation in iSE and SE yields as a function of the angle of incidence ϕ of the beam to a flat surface is similar, but not identical, as demonstrated by Fig. 4.8. A comparison of the electron-generated SE topographic yield curve as a function of the incident angle as compared with the iSE yield curves from H^+, He^+, Ne^+, and Ga^+ ions shows that the electron-generated SE variation as a function of the incidence angle is the smallest, while the iSE yield variation with angle is the highest for the heaviest ion. It can also be seen that for the heavier ions, the iSE yield actually reaches a maximum value after which it hints at falling back a little as the incident angle approaches closer to 90°. Such effects may not generally be of much significance except, for example, when line profiles across 3-dimensional objects are being recorded and analyzed for the purposes of metrology and control. In that case, then the actual chemistries of the beam and specimen could affect the apparent height and slope of features and careful calibration will be required (Fig. 4.9).

4.7 Ion Channeling Contrast

Channeling contrast is the name given to the variation in signal level observed from a crystalline material as a result of changes in the orientation of the incident ion beam with respect to the specimen (Fig. 4.10). While this phenomenon occurs for both ion and electron beams, the contrast level for ion can be as high as 30–40 % as compared to just 3–5 % in the electron beam case. When electrons interact with a crystal, they can produce "channeling patterns" which are geometric representations of the symmetry of the crystal lattice as viewed from the direction of the incident electron beam. When the crystal thickness is of the order

Fig. 4.9 iSE image of
diatoms. Field of view
20 mm and He$^+$ beam at
40 kV

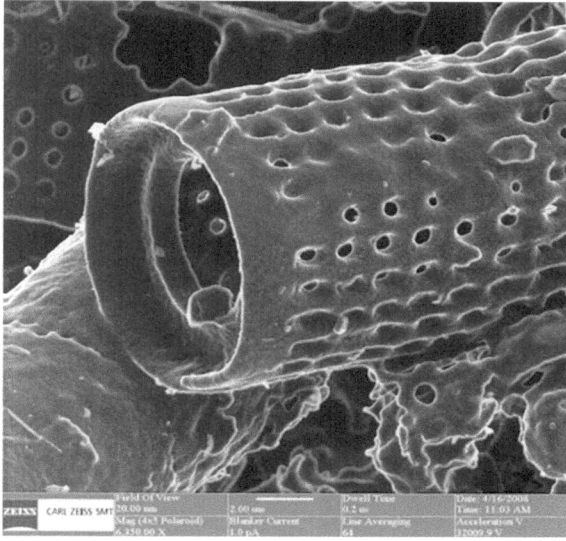

Fig. 4.10 Image of MoS$_2$
showing high levels of
crystallographic contrast
excited by incident He$^+$ ion
beam

of nanometers, then the channeling patterns are of high contrast and visibility, but
as thickness of the specimen increases, the crystallographic contrast becomes less
strong and on bulk samples is limited to just a few percent. In the electron beam
case, channeling contrast is best observed using the backscattered electron signal
mode.

There are significant differences between the response of electrons and ions to crystalline materials. The wavelength of ions is much shorter than that of electrons, and no patterns are formed. Instead, regions of varying crystallographic orientation show up as being of different brightness. If the material is tilted with respect to the beam direction, then the contrast remains unchanged until a critical angle, typically a few degrees in magnitude, is reached. Each crystallographic region then resets to a different brightness level. Even when using the usual iSE detector, channeling contrast can be as high as 40–50 % and so is one of the most frequently observed imaging effects. However, if the beam is allowed to scan the same area for long periods of time, then the channeling pattern may weaken or disappear as a result of the beam damage generated by the incident ions.

Ion channeling contrast responds not only to changes in the crystallographic orientation of a material but can also reveal individual crystalline defects such as dislocations and grain boundaries. It is also a sensitive indicator of the cleanliness and surface condition of the sample under examination because the contrast becomes weak, or disappears completely, if the specimen surface is not clean or if it has been mechanically damaged, for example by abrasive polishing. The first ion microscope, designed and constructed by Ricardo Levi-Setti at the University of Chicago, was in fact designed to display channeling contrast so as to determine the orientation and interfaces of minerals (Levi-Setti 1974). Recent work by Giannuzzi and Michael (2013) has provided a good introduction to interpreting these crystalline contrast effects.

4.8 Backscattered Ion Imaging

As noted in the previous sections, ion-induced secondary electrons (iSE) are a convenient and uniquely powerful tool for imaging in the HIM, but the interaction of the incident ions with the sample produces other signal types which can provide additional information to compliment the iSE data. For example, some fraction of the incident ion beam is backscattered, i.e., the ion is deflected through an angle greater than 90°, a process often called Rutherford backscattering (RBS) imaging (Rutherford 1911) (Fig. 4.11).

Any type of "backscattered detector" could be used in the HIM to exploit this mode of imaging, including an Everhart–Thornley (ET) detector operated with no applied high voltage bias. However, the ET option is not usually the optimum choice because that type of detector is physically small in order to reduce its capacitance and enhance its collection speed, and so when positioned to one side of the sample, it collects only a small fraction of the total available signal. Large "X-ray" solid-state detectors can be positioned directly above the specimen and so subtend a high solid angle for efficient signal collection, but the high capacitance

Fig. 4.11 Comparison of an iSE image (*left*) with a chemically sensitive RBI image (*right*) of copper mesh exposed to the atmosphere. Beam energy 36 kV and field of view 20 mm

of these devices restricts them to slow beam scan speeds. However, the recently introduced silicon drift detectors (SiDD) are likely to be of great value for RBS operation because these devices have low capacitance and a very high maximum counting speed. When the detector is used together with a modern high-speed X-ray multi-channel analyzer system, they can operate in a digital mode and so display and record not only conventional images but can also display the energy spectrum of the RBS signal, or perform filtered imaging within a selected energy window.

A plot of the RBS signal level as a function of the atomic number of the material examined (Fig. 4.12) shows that the detected signal does indeed generally increase with the atomic number of the target material, but it also tends to exhibit some oscillatory behavior which implies that the recorded signal level (i.e., the pixel brightness) is not monotonically proportional to the mean atomic number (Z value). The variations in the RBI signal versus atomic number signal occur at and around those atomic numbers for which the outer shell of electrons is complete. This behavior, however, means that the relationship between signal level and the atomic number Z of the sampled material is not unique nor single-valued. If the specimen contains multiple elements, then assuming that the RBI signal brightness is related to the averaged atomic number of the sample is probably an acceptable approximation, but the exact details will vary between different detector types such as microchannel plates (MCP) and silicon drift detectors (SiDD), so it is essential to do a calibration of the system using other similar, but better known, materials. In order to be able to apply the backscattered ion signal for reliable quantitative microanalysis, a very different procedure must be employed, and this is discussed later.

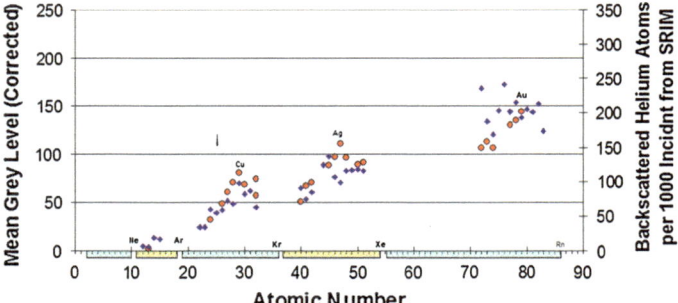

Fig. 4.12 Plot of RBI signal as a function of the atomic number of the target material

4.9 Scanning Transmission Ion Microscopy

As discussed earlier, the penetration depth of ions of 10 keV energy or more into a material is always significantly less than the range of electrons of the same energy and in the same material. However, if a specimen can be made sufficiently thin, then a significant fraction of the incident ion beam can still penetrate through the material and provide transmission ion images. The only changes required to the ion beam microscope are to provide some device which can hold the specimen to be examined and a detector beneath the specimen to collect the signals. Figure 4.13 shows an example of a basic scanning transmission adaptor (Woolf et al. 1972) which has been modified for use in an HIM. The specimen, typically a thin foil, or particles on a thin substrate, is mounted on a standard 3-mm-diameter grid of the type used in a conventional transmission microscope. Beneath the specimen, the beam path is shrunk to a slit about 0.25 mm in width which allows only those ions on, or close to, the center of the specimen holder to pass on down through the device so minimizing the number of scattered ions and enhancing the contrast. The transmitted ions then strike a gold-shadowed silicon wafer which is

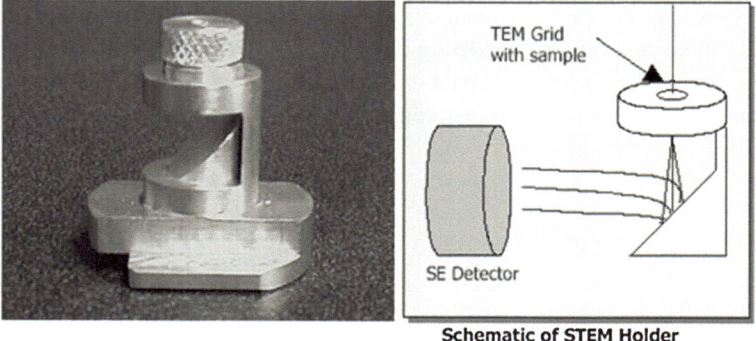

Fig. 4.13 Basic HIM adaptor for scanning transmission imaging

Fig. 4.14 Computed transmission of He$^+$ ions in silver (**a**) at 10 kV energy; (**b**) the corresponding range at 30 kV and the range at 100 keV

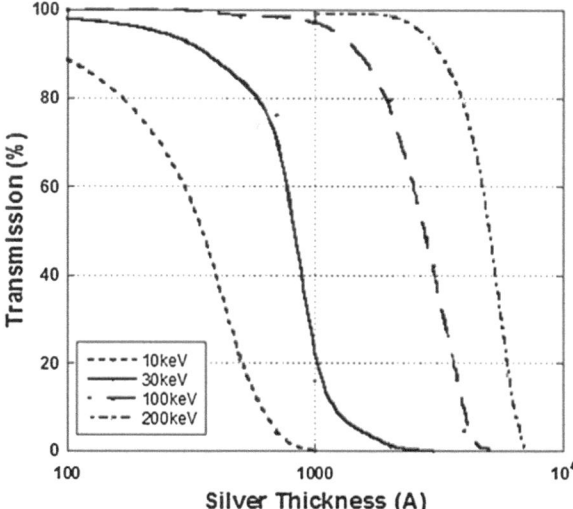

tilted at 45° to the vertical and oriented so as to face the ET, or any other, SE detector which finally collects the signal. This device, although simple, is both functional and surprisingly efficient as each transmitted ion which reaches the tilted silicon/Au target produces as many as 5 or 10 secondary electrons for delivery to the detector. More elaborate versions of this detector configuration (for example Notte et al. 2009) make it possible to simultaneously generate and record both bright and dark field images.

Because the stopping power of ions is much higher than that for electrons, specimens for observation in ion transmission mode must be significantly thinner than those used for electron beams. Calculations using the IONiSE Monte Carlo simulation (Ramachandra and Joy 2010) show (Fig. 4.14) that for a He$^+$ beam and a silver target at 10 keV, the bright field transmission signal falls by 50 % for a layer thickness of 30 nm and the resolution degrades, as a result of beam broadening, to about 5–10 nm. At 30 keV, the corresponding figure for an acceptable sample thickness would be 80 nm and about 250 nm at 100 keV. In materials of lower atomic number and density, such as MgO, then the usable range for comparison at 30 keV would be 75 nm and the beam broadening would be 3 nm. Although these estimated useable thickness values are much lower than those for comparable electron beams, they are still large enough to permit quality images to be obtained. The limited beam range associated with ion beams becomes a significant advantage when imaging materials such as graphene because then even a single sheet still produces good image contrast and detail in the transmission mode at 40 keV. Using a simple transmission adapter provides transmission images of thin films, such as carbon, which combine excellent resolution with high image contrast and a good signal-to-noise ratio. This in turn makes it possible to use lower values of incident beam current (typically less than 1 pA) so minimizing ion-induced damage. These images are very similar to those generated by a conventional SEM but with the benefits of a better signal-to-noise ratio, improved resolution, and an enhanced depth of field.

Observations of crystalline materials in the transmission ion mode produce images that look familiar when compared to transmission electron microscope images of similar materials. For example, Fig. 4.15 shows the image of an MgO cube resting on one edge. The alternating, periodic, bright, and dark bands can be recognized as being the thickness fringes (Hirsch et al. 1977; Williams and Carter 1996) observed in TEM images. From the dimensions of the cube, and a measurement of the fringe spacing, the extinction distance is calculated to be 15 nm, which is about 40 % smaller than the value for the same specimen when imaged with 40 keV electrons and is to be expected because of the stronger interactions associated with ions. Other images show clear evidence of dislocation structures, visible in the bright field (on axis) image as dark lines. When imaged by an electron beam, even small changes in the relative orientation of the specimen to the incident beam direction are sufficient to eliminate some areas of contrast and to reveal other new contrast features. However, when using an ion beam, the wavelength λ is so small in magnitude that the Ewald sphere (Hirsch et al. 1977) is effectively flat, and consequently, a large number of diffracted beams will be simultaneously excited at all orientations and the effects of beam tilt will become less dramatic. The appearance and behavior of such dislocation images can be matched to the predictions of the standard electron beam dynamical theory (Hirsch et al. 1977) although such a model represents a significant oversimplification of the phenomena involved. Further detailed study and analysis is clearly required before any definitive statement can be made.

Fig. 4.15 He$^+$ ion beam transmission images of MgO in water, displaying thickness fringes and dislocations

4.10 Detecting Ion Beam Signals

It is probable that for most applications, the device used to collect the iSE signal in their HIM will be the same "ET" detector that has been the choice for secondary electron imaging for more than fifty years since the publication of the classic original paper (Everhart and Thornley 1959). The details of the standard SEM "ET" detector are discussed in detail elsewhere (e.g., Goldstein et al. 2003) and will not be repeated here. It is important, however, that the performance of this, and all other, detectors should not be taken for granted but should be regularly tested and evaluated because the signal is at its weakest level as it initially reaches the detector. Signal information lost at that stage cannot subsequently be recovered. In particular, detector performance will inevitably degrade with time in an HIM as a consequence of the continuous impact of energetic ions on to the phosphor screen. The sensitivity of the ET detector, and indeed most other detectors, can degrade significantly within a period of just a few months as a result of its continuous exposure to energetic ion, so frequent checks are essential to maintaining the best performance at the most crucial point of the signal chain.

The efficiency of a detector is given by its "Detector Quantum Efficiency" (DQE) (Joy and Voelkl 1998) which is defined as

$$\mathrm{DQE} = (\mathrm{S/N})^2_{\mathrm{expt}}/(\mathrm{S/N})^2_{\mathrm{theory}} \tag{4.5}$$

where $\mathrm{S/N_{exp}}$ is the measured signal-to-noise ratio of the output signal from the detector and $\mathrm{S/N_{theory}}$ is the predicted signal-to-noise ratio for these operating conditions. A "perfect" detector would then have a DQE of unity and a non-functioning device a DQE of zero. Given that any detector can only occupy a limited fraction of the total space available in the specimen chamber, then experimental DQE values from detectors performing at a satisfactory level should be expected to fall in the range from 0.25 to 0.5. At the other extreme, a DQE value higher than about 0.5 would not, in general, be optimum as this would tend to reduce the effectiveness of iSE imaging by mixing, and so weakening, the light and shadow effects associated with topographic contrast. DQE measurements made on recently installed a number of SEM and HIM systems have produced measured DQE values that are only of the order of 0.1 even before any significant use that could damage the phosphor has occurred. If these values are really typical, then it indicates that a considerable fraction of the potential performance of which an ion microscope could be capable might actually already being wasted. On microscopes that may have already encountered significant use, a less-than-ideal DQE values might simply be the result of damage to the scintillator material caused by the use of excessively high beam currents, or because the detector has been in use for a period of a year or more, or even because the detector is located in an unsuitable position in the chamber. In that case, once a measurement has been taken, it is a simple matter to refurbish the faulty detector and restore it to the original level of performance.

4.11 Practical Issues in HIM with a GFIS

When using an SEM, it is always recognized to be good practice to keep the working distance (WD), i.e., the spacing between the sample and the final lens of the instrument as short as possible is consistent with the need to allow detectors to properly view the regions of interest because this reduces the severity of the electron optical aberrations that result when lens excitation is weakened. Reducing the working distance in the HIM is not so critical because the aberrations of the ion optics are much less serious in magnitude because the beam convergence angle itself is so small. Consequently, a working distance of between 3 and 5 mm should be optimum for most purposes. However, operating at a working distance of 10 mm or greater could still result in high image resolution and an excellent depth of field while providing a wider field of view and eliminating the risk of having a bulky sample cause damage to the lens.

SEM operators are used to adjusting the energy of the incident beam so as to optimize different facets of performance. For example, a high beam energy will chosen for obtaining the best possible resolution, or when significant beam penetration into the specimen is required, and for microanalysis. But lower energies would be selected to image surface detail, or for observations of samples that are fragile, or of low electrical conductivity. When using the HIM, however, the situation is different. As discussed earlier, the landing beam energy can be set anywhere between about around 35 keV and lower. But because the iSE yield and the ion gun brightness both become less as the beam energy is reduced, the imaging performance of the microscope will tend to degrade without gaining any compensating benefit except a further reduction in beam penetration. Under most conditions, therefore the HIM should be operated at the highest possible energy.

The HIM is likely to be more susceptible to vibration and other mechanical disturbances than an SEM because the ion source is only demagnified by a modest factor of 3x–5x at the specimen position as compared with factors or 100x or more typically encountered for an SEM. The weight of the ORION microscope is, by design, substantial enough to help minimize this effect, but placing matting or foam around the walls of the laboratory will help to minimize noise and improve the quality of images from the instrument. On the other hand, the HIM is nearly impervious to the effects of magnetic fields from the power wiring and other pieces of laboratory equipment, but the 50 and 60 Hz power supply lines can still cause image interference problems as a result of ground loops and poor grounding and these must be eliminated if optimum performance is to be achieved.

In addition, it is worth noting that the beam spot size at the specimen is always likely to be only a few nanometers or less in diameter (Ward et al. 2006). When imaging at the nanometer level, this is, of course, essential, but it should be remembered that anytime the imaging field of view is enlarged (i.e., going to lower magnification operation) the pixel size in the image may become very large by comparison with the size of ion beam spot itself. As a result, the gray level of each pixel in the image is determined by the probe size and is not necessarily representative of the gross area of the much larger pixel. Although this effect can be minimized by increasing the spot size, the HIM is not capable of generating large (micrometer or larger size) beams, so the number of image pixels should be increased where possible to help minimize the spurious artifacts which might otherwise become evident.

An obvious final question is "how good could the HIM ultimately become?" The optical performance of the HIM is already guaranteed to be highly competitive because the usual electron optical problems of spherical and chromatic aberration and the shallow imaging depth of field are already substantially absent. In addition, the beam spot size is in the low subnanometer range and can readily be reduced still further in size, the signal-to-noise ratio is superior to that of the SEM, and still better specifications in all these parameters could be obtained by increasing the ion beam energy to the 100 kV range or higher. Set against this, there are some practical considerations. For example, reproducing a particular beam voltage is difficult, because the actual landing energy varies from one sample to the next, so direct comparisons will more difficult to perform, and some sample damage is inevitable. The option of being able to change the ion beam is important in minimizing some of these concerns, but the time required to switch between different sources may be some lengthy and, as discussed in a later chapter, it is not yet certain that the same GFIS can be persuaded to perform equally well for all gases.

Chapter 5
Charging and Damage

A major concern in both scanning electron and scanning ion microscopy is that of sample charging, but strategies to eliminate this problem are available. In the SEM, this occurs when electrons impact nonconducting samples. Charging occurs because electrons neither be created nor be destroyed, so currents at a point must sum to zero. For a material irradiated with electrons,

$$I_B = I_B(\eta + \delta) + I_{sc} + \mathrm{d}Q/\mathrm{d}t \tag{5.1}$$

where η and δ are the BSE and SE yields, respectively, and Q is the static charge on the specimen at some time t, I_B is the incident beam current, and I_{sc} is the specimen current flowing to ground. If the sample is a conductor then it will not charge and so Q will be equal to zero all times. In this case, and when operating at high beam energies where electron yields are small, any excess current will flow to earth as specimen current I_{SC}. At low beam energies where signal yields are higher then current flows from earth to the sample to make up the deficit, the charge is always balanced and so stable imaging is possible. But when the material under observation is an insulator, then the specimen current I_{SC} is zero. So if the specimen is not to acquire a static charge, then $\mathrm{d}Q/\mathrm{d}t$ must be held constant at zero. This can be achieved if the incident current is exactly balanced by the emitted beam current:

$$I_B = I_B(\eta + \delta) \text{ i.e. if } \eta + \delta = 1 \tag{5.2}$$

and this condition represents a dynamic charge balance. Otherwise, if $(\eta + \delta) < 1$, then negative charging will occur, or if $(\eta + \delta) > 1$, then the charge buildup will be positive.

The situation for either helium or any other of the ion beams available is different (Fig. 5.1). Consider an ion beam impinging on a piece of nonconducting material. The incident ion beam is positively charged and, as seen earlier, iSE yields are typically greater than unity so each incident ion will result in the ejection of one or more negatively charged secondary electron from the specimen. The incident ion will then eventually come to rest in the sample, depositing a residual amount of positive charge into the material. So the impact of ions on a

D. C. Joy, *Helium Ion Microscopy*, SpringerBriefs in Materials, DOI: 10.1007/978-1-4614-8660-2_5, © David C. Joy 2013

nonconducting specimen results in the ejection of negative charge and deposition of positively charged. As a result when insulating materials are irradiated by an ion beam, they will always acquire a net positive charge. The SE image will now show a dark area in the region scanned by the beam indicating that secondary electrons are unable to leave the sample and go to the detector because of the buildup of positive charge on the surface. Some scheme for eliminating, or at least minimizing, charge buildup on specimens during observation, is a necessity for HIM operation.

Several different approaches for imaging specimens that are charging under the ion beam are in use. For example, with the addition of an electron gun into the specimen chamber, the sample is periodically flooded by a beam of electrons with an energy of a few hundred electron volts or less. At the end of each line scan, the ion beam is blanker or switched off and the electron beam is turned on. The output of the electron gun is dispersed over an area centered on the optic axis of the microscope and extending for a few millimeters in every direction. After a dwell period of a few milliseconds or so, the electron flood gun is switched off, the ion beam is switched back, and the next line of the image is scanned. This process is continuously repeated to assure stable imaging and operating conditions, and the landing energy, and the duty cycle, of the electron flood gun can be varied to achieve the best overall result. This procedure works well and, because it is fully automated, requires little or no operator intervention. However, it is also somewhat cumbersome, and the electron flood gun is itself supplying additional, and potentially damaging, charge to the specimen which would be especially undesirable if the material under examination was to be highly sensitive to radiation.

An alternative and increasingly popular approach to managing ion-beam-induced charging is to inject a gas into the specimen chamber. A typical arrangement uses a small pipette, of the order of 100 μm or so in inner diameter, to carry air—or some other gases as desired—and terminating in a nozzle aimed directly toward to the specimen, but positioned a centimeter or so away from it. The injected gas forms a plume which flows across the surface of the specimen and

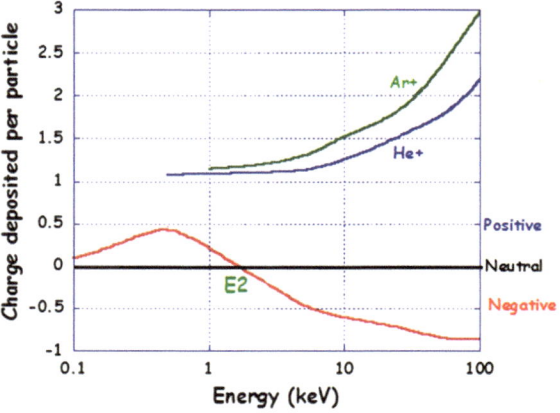

Fig. 5.1 Charging of insulators by an electron beam (red) or an ion beam (green and blue)

which becomes ionized by the incident beam as it travels. These ions and electrons then migrate toward oppositely charged regions of the specimen, reducing and then ultimately erasing that charge. More elaborate systems use one or more precision injectors each with independent control.

The gas injection procedure works well, is self-controlling, and, unlike the previous method, does not risk causing additional damage to the specimen as the result of depositing an additional source of ions onto the surface. However, each time the operator addresses a new field of view, it will take a few seconds for the image of the newly selected region to become stabilized and ready again for observation.

The ability to control or, if possible, eliminate sample charging is especially important when using ion beams because otherwise large electric fields can be generated even if the specimen and its image might seem to be stable. The effect of charging is to alter the landing energy of the ion beam which in turn will affect the dimensions of the scanned field of view and so give rise to image distortion. Effective, but nonintrusive, charge control is therefore essential.

Chapter 6
Microanalysis with HIM

6.1 Electrons, Ions, and X-rays

For many users, the most important application of an SEM is its ability to identify the chemical composition of a specimen. Energy dispersive spectroscopy (EDS) of the X-rays fluoresced from samples of interest by the incident electron beam provides chemical microanalysis combining unparalleled sensitivity together with high spatial resolution for elements across the entire periodic table. This technique would therefore also be the automatic first choice for microanalysis when using ion beams if it were a viable option.

Unfortunately this is not, in general, going to work for ion beams because the condition that must be satisfied in order for characteristic X-rays to be excited by incoming charged particles is that the velocity of the incoming ion or electron must equal or exceed the velocity of the orbiting electron which is to be ionized. [Joy et al. 2007]. When the incident particle is also an electron, these two energies are numerically identical when their velocities are the same. But when the incoming particle is a helium ion—and so is some 7,300x more massive than an electron—the kinetic energy of the ion must increase by a factor of 7,300x times in order to achieve the velocity match which will make ionization possible. Thus, for example, to excite Cu Kα, radiation requires helium ions with energy of about 70 MeV. Given that current HIM systems are limited to a maximum beam energy which is a factor of 1,000 time or smaller than this figure, there is clearly a need for some alternative approach if chemical analysis is required.

One approach to ion-beam-based microanalysis is to make use of the Rutherford backscattered ions (RBI) that are generated. But rather than relying on the magnitude of the ion yield to distinguish materials, as discussed previously, it is the energy of the backscattered ions that is measured here instead. For a given ion type and energy E_{inc}, the energy E_{iBS} of a back scattered ion is a function of the angle θ, relative to the incident beam direction, at which the ion is collected as well as the atomic mass M_1 of the incident ion and M_2 the atomic mass of the target material. These parameters are linked by the equation:

D. C. Joy, *Helium Ion Microscopy*, SpringerBriefs in Materials,
DOI: 10.1007/978-1-4614-8660-2_6, © David C. Joy 2013

$$(E_{\text{iBS}}/E_{\text{inc}}) = \left\{ \left(\left(1 - (M_1 . \sin\theta/M_2)^2\right)^{1/2} + (M_1 . \cos\theta/M_2)/(1 + M_2/M_1)^2 \right\} \right.$$

(6.1)

An energy selective detector, such as a solid-state diode, is positioned at some selected "take-off" angle θ relative to the incident ion beam direction. The mass M_1 and the energy of the incident ion E_{inc} are known, and the scattered ion energy E_{inc} is measured by the detector. The mass M_2 of the scattered ion can then be determined from Eq. (6.1) (Rutherford 1911). Figure (6.1) plots the relationship between the energy of the incident and scattered ions $E_{\text{iB}\Sigma}/E_{\text{iv}\chi}$ and the ion mass M_2 for a take-off angle θ of 90°, showing that each element in the periodic table has its own unique ion energy E_{iBS} as compared to the incident energy. With a stable detector and modern high-speed electronics, such scattered ions can be collected, automatically identified, and their mass values determined and stored at a rate of several thousands per second. The peak to background ratio of these signal can be quite high so the dynamic range of this approach is good, allowing major, minor, and trace elements to all be detected given a suitable detector.

While RBS analysis is a valuable analytical tool, it does suffer from some inherent, and ultimately intractable, problems. First, as is evident from the graph in Fig. (6.1), the spacing between successive element peaks in the period table decreases rapidly as their atomic number increases. The detector resolution, on the other hand, is a constant fraction of the energy of the incoming ion so the ability of the detector to separate and identify individual peaks will degrade rapidly with increasing atomic number, making reliable identification of heavy elements increasingly difficult and unreliable.

In addition ions, unlike X-rays, undergo both elastic and inelastic collisions within the sample and so will deposit some fraction of their energy as they travel

Fig. 6.1 Relationship between the energy of the incident and scattered ions, and the ion mass for a 90° take-off angle

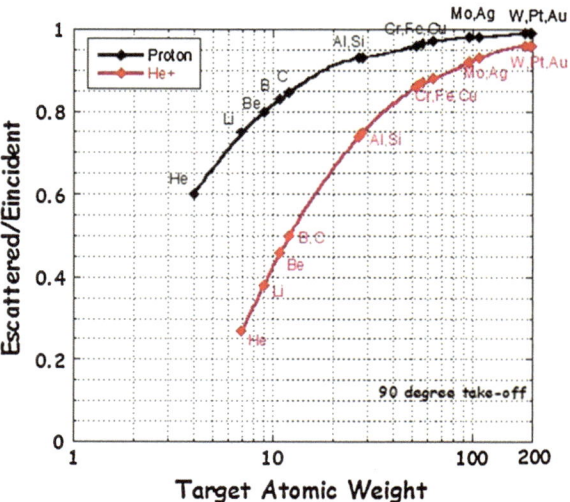

Fig. 6.2 Effect of variations in sample thickness on RBI analysis data

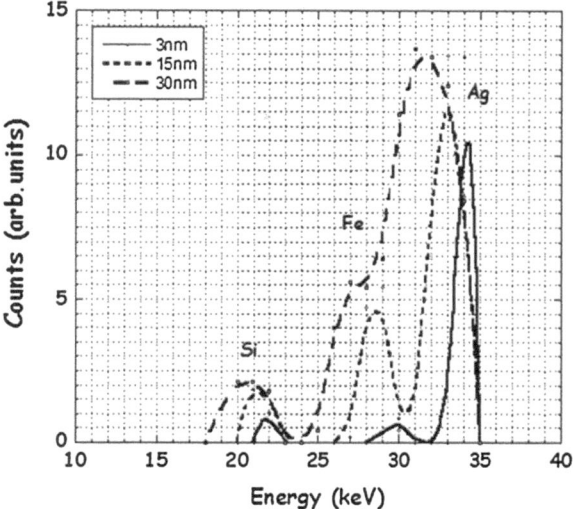

through the material. As a result, the spectrum that finally reaches the detector will be a function of the thickness of the specimen, as well as of its composition, and of the beam energy. The effects of this are visible in Fig. (6.2) which uses the SIMNRA code to simulate the predicted RBS spectrum, from a sample containing equal amounts of Si, Fe, and Ag, and for three different film thicknesses, and using a 35-keV He$^+$ beam. When the sample is just 3 nm thick, the Si, Fe, and Ag peaks are clearly separated, but as the thickness is increased first to 15 nm, and then to 30 nm, the mean energy of each of the elemental peaks decreases in magnitude, and the peaks themselves become broader and ultimately may merge into their neighbors. The RBS technique will therefore be least useful on samples which are thick and/or contain high-atomic number materials, and when the beam energy is low.

6.2   Time of Flight: Secondary Ion Mass Spectrometry

Conventional X-ray chemical microanalysis in the SEM can provide information about the elements present in the sample, but although such information is useful, and often even critical, it provides no information about the actual state of these elements. For example, the presence of carbon could indicate the presence of just the pure element itself, but it could also be simply one of the components of a complex polymer or biological sample. There is therefore a continuing need to be able to identify fragments of complex materials, rather than just pure elements. This requirement might possibly be satisfied by means of secondary ion mass spectroscopy (SIMS).

If a beam of ions (positive or negative), with energies in the keV range is scanned across a material, then the impact of the ions on to the specimen surface will result in the removal and ionization of not only some individual atoms, but also of clusters of atoms from the top two or three atomic layers. There are several different ways in which this beam-generated flux of charged particles could be analyzed to provide information about both individual atoms and more complex fragments, but the method that seems to be best suited for ion beam microscopy is time-of-flight secondary ion mass spectrometry (TOF-SIMS) as developed by Benninhoven et al. (1989) and others.

In this technique, individual ions and ion clusters are distinguished one from another, and then classified, by determining their "mass-to-charge" ratio. The energy, and hence the velocity, of the charged ions liberated from some point on the specimen surface illuminated by the beam is very low and differs randomly from one ion to the next. A large, fixed, potential difference of magnitude "U" is applied and maintained between the specimen and the target. This potential is chosen to be high enough in value to ensure that all the emitted ions from the sample will reach the target with the same final energy. If the mass of an emitted fragment is "m", its charge is "q", and its velocity when it strikes the target is "v", then by conservation of energy, and assuming the initial energy of the fragment was negligible, the final particle energy E is given as:

$$E = {}^1\!/_2 . mv^2 = q.U \tag{6.2}$$

If the path length from the position from which the ion emerged from the specimen surface to the detector is "d", then the transit time "t" required for the complete journey from the specimen to the target will be

$$t = d/v \tag{6.3}$$

Eliminating the velocity (v) from these two equations gives the result

$$t = (d/\sqrt{2U}) . (\sqrt{(m/q)} \tag{6.4}$$

The first term contains the physical parameters which describe the length of the flight tube and the voltage applied to it. These parameters are constant for a given system. The second term contains, in units of the mass-to-charge ratio of the ions, the desired chemical information about the specimen. Analysis of the data is therefore straightforward and with current electronics, hundreds or thousands of ions can be detected and classified per second. For the time of flight technique to be useful as a complement to the imaging capability of the HIM, it is necessary for the incident ion beam to be able to dislodge the material on the top surface layer of the specimen. This is most easily achieved by using a heavy-ion beam. For example, the so-called dual-beam instruments, which combine a scanning electron microscope (SEM) and a focused ion beam system, invariably use Ga^+ as their ion beam because of its ability to sputter the surface, but rely on the electron beam from the SEM to perform imaging with less damage.

Fig. 6.3 Sputter yield data
for H⁺ and He⁺

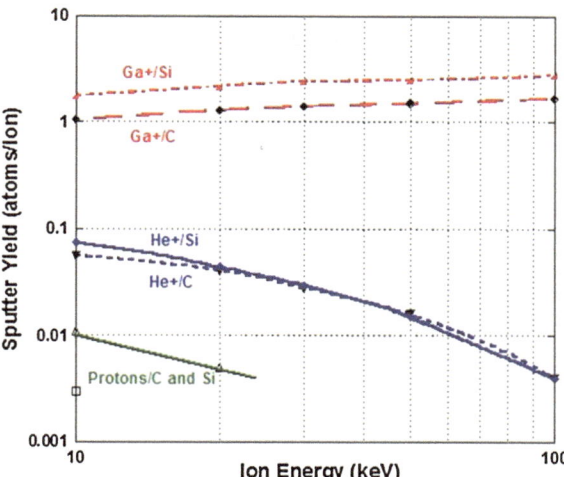

The He⁺ ions used in the HIM are able to provide high-resolution imaging but is only capable of sputtering most elements and materials at a rate which is too low to be practical for meaningful microanalysis (Fig. 6.3). The ORION HIM solves this problem by additionally providing a neon (Ne⁺) beam. This gives a sputter rate which can reach about 40 % of the corresponding value for Ga⁺ and is ten times higher than that for He⁺. Although the sputter rate is lower than for gallium, the neon beam does not implant any residual ions into the irradiated volume and so provides cleaner spectra. The yield of ions available for analysis can be further enhanced by providing chemical assistance in the form of oxygen, or a variety of other specially formulated chemical compounds, which are injected into the sample chamber. This ability of ion beams to break up and remove surface material is an important step toward using an ion microscope for "time-of-flight mass spectrometry". Other gases such as Ar⁺, which are less heavier than Ga⁺, could also be good for milling but will penetrate to greater depths and so initiate a larger mass loss. The "time of flight" of sputtered particles, from the initial ion impact to their arrival and analysis at the detector, is of the order of nanoseconds and with a typical incident ion current of 10 pA, and for atomic mass range up to 500 amu, the average count rate would be expected to be several thousands per second and would offer a mass-to-charge resolution, approaching a few parts in a thousand.

If the incident ion beam continues to raster the specimen while chemical data is being collected, then a two-dimensional chemical map of the surface layers can be generated. If the same area is scanned repeatedly, then successive chemical images will be coming from regions which are progressively deeper beneath the original surface from which data were obtained. TOFSIMS could therefore become a means of true, three-dimensional, high-sensitivity, high-performance microanalysis.

Although some demonstrations of TOF-SIMS operation have recently been made using Ga^+ sources attached to a standard dual-beam FIB system, there has been little published information on the performance that might be achieved when using either Ne^+ or other relatively light beams. However, the potential of this technology to expand the boundaries of microanalysis from basic elemental identification to the recognition and quantitative analysis of complex compounds will certainly be of increasing importance.

Chapter 7
Ion-Generated Damage

It has to be expected that both ions and electrons will damage specimens under examination to a greater or lesser degree. Electrons are low in mass but can travel at velocities which are a significant fraction of the speed of light. In general, electrons do not significantly damage metallic or inorganic specimens, but even relatively low doses of electrons can be expected to chemically alter or destroy organic materials such as polymers and biological samples. The threshold dose above which damage can be expected is typically as little as one to two electrons per square angstrom and may result in mass loss of the target, or to the removal of a high fraction of certain chemical components such as hydrogen or phosphorus. Knock-on damage, in which an incident electron releases an atom from its surrounding so leading to large-scale damage of crystals, depends directly on beam energy, so, for example, the damage threshold for carbon (atomic number 6) is 85 keV, but is 240 keV for Si (atomic number 14).

Ions are at a minimum from $5,000\times$ to $500,000\times$ times heavier than electrons and so inflict significantly more damage. In addition, the nature of ion damage is more varied in form than that for electrons (Benninghoven et al. 1987) because ions can generate point defects such as vacancies and interstitials within the materials that they attack. The effects of such damage depends on the binding energy of atoms in crystal lattice, and on the displacement which they undergo. Typically, the displacement becomes negligible for energies above 100 keV. However, the sputter yield rate at such energies is still high enough—typically 0.1 atoms/per ion at 30 keV—to allow Helium ions to pattern, materials such as graphene sheets. A study by Livengood et al. (2006) showed that, while the damage caused to wafer silicon by 30 kV He^+ ions was negligible, at higher energies and/or when subjected to significantly larger beam doses, even materials such as gold nanospheres and carbon nanotubes could be damaged by the He^+ beam. Consequently, the absence of significant damage at one energy and dose cannot always be taken as being evidence of a similar outcome at some higher energy. On the other hand for those materials which can readily be sputtered, He^+ ions are ideally suited for patterning as the ions suffer little lateral scatter and so are capable, for example, of precisely cutting holes, or fabricating slots, no more than a few nanometers in diameter and spacing.

D. C. Joy, *Helium Ion Microscopy*, SpringerBriefs in Materials, DOI: 10.1007/978-1-4614-8660-2_7, © David C. Joy 2013

Chapter 8
Working with Other Ion beams

A feature of the GFIS ion source is that every aspect of its operation and behavior—from its imaging resolution, the energy range over which it operates, the efficiency of signal production, and the damage it does to the materials that it examines—is ultimately affected by the choice of imaging gas. Ideally, the same source could rapidly be reconfigured to select and generate any one of a number of different ion beams. Because each type of ion has its own strengths and weaknesses, this feature would add substantially to the utility of the ion microscope.

For example, the unique property of the lightest option—a proton beam—is that its operating performance combines some of the best features of both He^+ ion and conventional electron beam instruments. The secondary electron yields generated by protons are generally comparable in magnitude with those from a He^+ beam (Fig. 8.1a, b) but have the advantage that the iSE yield conveniently reaches its peak value for an energy of only about 50 keV rather than at 800 keV or more required for a comparable helium beam. In addition, the severity of beam-induced damage is significantly reduced in comparison to that from He^+ and other heavier gases. The most immediate problem is that hydrogen is highly flammable, and any arcing or electrical discharge in the emitter tip region could result in significantly and potentially highly dangerous discharges within the gun chamber. Nevertheless, the perceived benefits have been sufficiently exciting to have resulted in the construction of a number of proton microscopes.

The use of medium-heavy gases such as argon (Ar^+) or neon (Ne^+) is now attracting much attention because these should combine some of the best features of both light and heavy ions. Although adequate iSE yields can still be generated even at beam energies as low as 30 keV (Fig. 8.1a, b), achieving the higher currents required to improve imaging performance will ultimately require energies of the order of 1,000 keV or higher. At such elevated levels, the beam probe size could then be expected to be significantly smaller than that from either H^+ or He^+ systems because the ion wavelength, which depends inversely on the particle energy, will be very much reduced.

In practice, however, a number of problems will first need to be solved before achieving such a level of performance becomes a routine proposition. For example, in the case of Ne ions, the energy spread of the beam is about 1 eV or more.

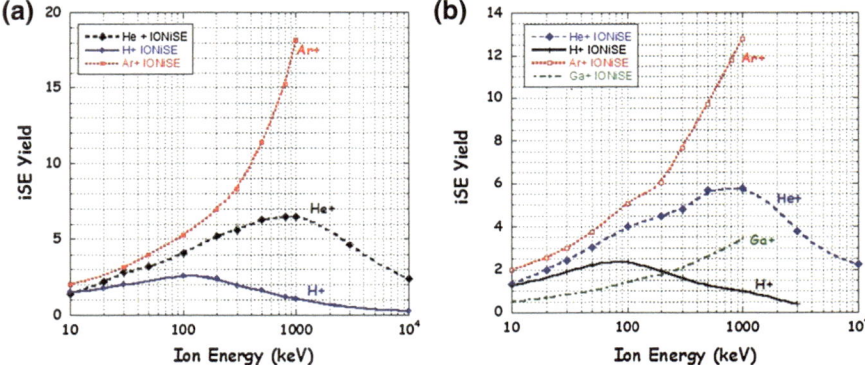

Fig. 8.1 The yield of secondary electrons depends on the target and on the energy and mass of the incident ion; **a** Predicted iSE yields from a silver sample irradiated with H$^+$, He$^+$, and Ar$^+$ ions as a function of beam energ; **b** Predicted iSE yield from a platinum sample irradiated wit H$^+$, He$^+$,Ar$^+$, and Ga$^+$ ions as a function of energy

This is greater than the values for either He$^+$ or H$^+$ and might limit the attainable resolution and reduce the contrast. Commercially available neon is actually a mixture of two isotopes (^{22}Ne at 9.25 % and ^{20}Ne at 90.84 %), and these components of the beam will not be brought to a focus to the same position, leading to a further loss of resolution and contrast. In addition, there are still other issues which are need to be considered, for example, Ne$^+$ beams have a tendency to exhibit brightness fluctuations with time scales varying from 10 ms to 10 s. Significant fluctuations in Ne$^+$ emission are also observed when monitored over a timescale of several hours, and the beam may suddenly disappear after running safely for a period of several hours. Remedies to minimize or eliminate these problems are under development and include improving the gun and its pumping speed, improving the gas box, studying the consequences of running neon at 90 % of the He$^+$ BIV energy, and further optimizing the tip building process. In any event, it is clear that much research is still required to produce a variety of high-performance ion beams with reliable long-term performance.

The most widely used ion for heavy-duty material processing is Ga, because it is readily available, and it is heavy and so is highly efficient at removing material. However, gallium, unlike the lower atomic number ion precursors, can become implanted in the target rather than escaping into the gas phase, which will cause chemical contamination as well as damage and may lead to the formation of gallium compounds in the target. Several different types of gun design have been successfully used with Gallium to generate ion beams with a spot diameter below 100 nm, the limit ultimately being determined by the energy spread inherent in the beam, and by the required current which can be of the order of 1 nA for energies of 50 keV. Consequently, a gallium source can sputter and cut materials very rapidly and so is widely used to machine or repair materials and to prepare samples for examination. It is not, however, an optimum choice for high-performance imaging. The drawbacks of using a Ga^+ source include the fact that it deposits gallium on, and possibly in, every surface and crevice in the specimen chamber, contaminating the other materials placed in the unit unless great care is taken.

Chapter 9
Patterning and Nanofabrication

As seen above, ion beams can rapidly remove material from a specimen placed in the HIM. Depending on the proposed application, this could then be considered either as damage—and so be undesirable—or as a unique tool to pattern material. The use of Ga^+ beams for thinning or cross sectioning materials prior to examination in a transmission electron microscope is well known and in widespread use. As noted earlier, less known is the fact that light ions such as He^+ can also remove material from a surface, although at a much reduced rate, providing a method to shape, mark, and pattern materials on nanoscale. The typical rate at which material can be removed by sputtering by a He^+ beam offering 1pA emission is only one or two atomic layers per second over an area of a few nm^2. However, this still permits the He^+ beam to lay down a pattern, consisting of an array of dots or simple linear structures, on silicon or even on advanced materials as graphene. The ORION NF helium ion microscope comes complete with a simple pattern generator for such applications as these, but a more advanced system offering more options and faster writing speeds, and pattern generation automation will be required for serious use.

In addition to patterning, which removes material from a target, the direct deposition of selected materials on to the target is also possible. As discussed earlier, precision gas injectors can be used to direct a fine gas plume, usually of the order of 100 μm or so in diameter, of an active precursor material toward the sample. In areas where there is no ion beam present, the injected material will be deposited on the sample surface but then rapidly vaporize and be harmlessly pumped away. However, if the ion beam is directed into a region within the precursor stream, then this will result in the deposition of material at the point of contact of the beam with the surface. At all the other surrounding areas within the plume, the precursor will again vaporize and be removed by the chamber pump. This technology makes it possible to rapidly fabricate nanoscale, three-dimensional structures in a competitively short time. In addition, if more than one injector is available, then the chemical composition of the combined deposit can be varied to produce a variety of chemistry varying with both depth and horizontal position.

D. C. Joy, *Helium Ion Microscopy*, SpringerBriefs in Materials,
DOI: 10.1007/978-1-4614-8660-2_9, © David C. Joy 2013

Conclusion

Although clearly there is still much development and fundamental research remaining to be done, it is evident that ion beams and ion beam microscopy are no longer a novelty, or restricted to just mundane tasks, but are a valuable new technology with great promise for imaging, as well as for nanofabrication, and the analytical sciences. These capabilities will profoundly affect both the physical and the biological sciences over the coming next few years. In particular, if reliable operation can be guaranteed at beam energies above about 100 kV, then light ion microscopy will ultimately be capable of demonstrating atomic level resolution even on bulk materials, and microanalysis at the atomic level using TOF-SIMS or some other advanced technique will revolutionize the meaning of analytical microscopy. Ion beams will be "THE" microscopy for the twenty-first century!

D. C. Joy, *Helium Ion Microscopy*, SpringerBriefs in Materials,
DOI: 10.1007/978-1-4614-8660-2, © David C. Joy 2013

Appendix

iSE yields and IONiSE parameters for He^+ excitation of elements and compounds

Element/Material	iSE yield			IONiSE parameters	
	10 keV	30 keV	50 keV	λ(Ang)	ε(eV)
Li	0.69	1.20	1.51	25	40
Be	0.7	1.25	1.47	9.3	65
C	0.55	1.2	2.6	10.0	59
Mg	0.65	1.3	2.55	10.8	41
Al	0.63	1.17	1.51	12.0	40
Si	0.29	0.63	0.75	11	100
Ti	0.6	1.1	1.6	8.5	65
Cr	0.8	1.2	1.7	7.5	70
Mn	0.8	1.25	1.7	8	65
Fe	0.85	1.3	1.7	9	75
Co	0.85	1.45	1.75	8	63
Ni	0.95	1.50	1.9	8	69
Cu	0.76	1.27	1.51	9	63
Zn	0.9	1.4	1.8	12	60
Ge	0.8	1.2	1.37	9	60
Zr	0.85	1.5	1.85	8.5	75
Nb	0.9	1.65	2.0	8	75
Ag	1.5	2.74	3.2	10	50
Cd	1.35	2.7	2.9	10	45
In	1.2	2.15	2.45	9.5	45
Sn	1.2	2.0	2.6	9.5	50
Ta	1.0	1.9	2.3	9	80
W	1.15	2.1	2.6	7.5	78
Pt	1.25	2.45	3.2	7.5	60
Au	1.42	2.5	3.0	9.2	67
Pb	1.2	2.1	2.6	15	90
MoS_2	0.45	0.79	1.06	10	1
Stainless Steel	0.59	1.28	1.5	8	75

D. C. Joy, *Helium Ion Microscopy*, SpringerBriefs in Materials,
DOI: 10.1007/978-1-4614-8660-2, © David C. Joy 2013

The λ and ε parameters shown here can, when used together with the IONiSE program (available from http://djoy@utk.edu), predict the iSE yield of these materials as a function of the incident He+ beam energy and their angle of incidence to the beam. Although there are a number of published works containing experimental iSE yield measurements, the reliability of such data cannot be taken for granted as the cleanliness and the smoothness of the surface of the material under test can affect the apparent iSE yield by a significant amount. At present, iSE yield data that are believed to be reliable are only available for the limited number of elements listed plus the two materials—so much work still needs to be done.

For those materials included in the list above, the same and parameters when used together with IONiSE can also predict the iSE yield resulting from H^+, Ar^+, Ne^+, or Ga^+ ion beams even though there is at this time little corresponding experimental data. As shown (Fig. 8.1), the magnitude and behavior of the iSE yield with ion type and energy vary greatly from one type of beam to another in the same energy range, so the ability to generate and employ several different ion beams is of value.

(

Bibliography

Cited Papers

Bethe H (1941) The generation of secondary electrons. Phys Rev 59:940–942

Beninhoven A, Rudenauer FG, Werner HW (1987) Secondary ion mass spectrometry. Wiley: NY, pp 339–664

Chen P(2010), van Veldhoven E, Sanford CA, Salemink H, Maas D, Smith DA, Rack PD, Alkemade PFA Nanopillar growth by focused helium ion beam induced deposition. Nanotechnology 21:455302 (1–7)

Crewe AV, Eggenberger EN, Wall J, Welter LM (1968) A field emission gun microscope. Rev Sci Instrum 39:576–580

Dapore M (2011) Secondary electron emission calculations performed using two different Monte Carlo strategies. Nucl Instr Methods Phys Res B269:1668–1671

Everhart TE, Thornley RFM (1960) A detector for secondary electrons. J Sci Instrum 37:246–250

Fresnel A (1826) Memoires de l'Academie des. Science 5:475

Giannuzzi LA, Stevie FA (eds) (2005) Introduction to focused Ion beams: instrumentation, theory, technique, and practice. Springer, NY

Giannuzzi LA, Michael JR (2013) Comparison of channeling contrast between ion and electron images. Micros Microanal 19:344–349

Goldstein JG, et al (2003) Scanning electron microscopy and microanalysis. SEMXM 3rd edn

Hirsch PB, Howie A, Nicholson RB, et al (1977) Electron microscopy of thin crystals. R J Krieger, NY, pp 208–221 (Chapter 9)

Joy DC (2011) Scanning ion beam microscopy and metrology. A.I.P. Conf Proc 1395:80–83

Joy DC, Meyer HM, Bolorizadeh M, Lin Y, Newbury DE (2007) On the production of X-rays by low energy ion beams. Scanning 29:1–5

Joy DC, Voelkl E (1998) Quantifying SEM resolution and performance. Micros Microanal 4(supp2):276–277

Kanaya K, Okayama S (1972) A formula for electron range J Phys D Appl Phys 5:43–45

Levi-Setti R (1974) Proton scanning microscopy. Scan Electr Micros I:535–551

Livengood R (2007) HIM invasiveness and the imaging of semiconductors. J Vac Sci Tech 25:2547–2552

Morgan J, Notte J, Hill R, Ward BW (2006) Helium ion microscopy. Micros Today 14:24–31

Muller E, Bahadur K (1955) Field ionization of gases at a metal surface and the resolution of the field ion microscope. Phys Rev 102:624–627

Muller E, Tsong T (1993) Emission properties of electron point sources. Ultramicroscopy 50:57–64

Notte N IV, Hill R, McVey SM, et al (2009) Diffraction imaging in a He+ ion scanning transmission microscope. Micros Microanal 15:113–115

D. C. Joy, *Helium Ion Microscopy*, SpringerBriefs in Materials,
DOI: 10.1007/978-1-4614-8660-2, © David C. Joy 2013

Orloff J, Swanson LW (1977) A scanning microscope with a field ionization source. In: Proceedings of 10th annual SEM symposium on scanning electron microscopy, vol 1. IIT Research Institute, Chicago, pp 57–62

Ramachandra R, Griffin B, Joy DC (2009) A model of secondary electron imaging in the helium Ion scanning microscope. Ultramicros 109:748–757 (A current copy of the ion database associated with this model is available from djoy@utk.edu)

Rutherford E (1911) The scattering of alpha and beta particles by matter and the structure of the atom. Taylor and Francis, London, pp 688–700

SIMNRA. The current version of the SIMNRA program, developed by M.Mayer can be downloaded from www.rzg.mpg.de/vmam

Williams DB, Carter CB (1996) Transmission electron microscopy, vol 2. Plenum Press: NY, pp 191–199

Woolf RJ, Joy DC, Tansley DW (1972) A transmission stage for the SEM. J Phys E Sci Instrum 5:230–234

Young T (1803) An examination of lenses. Phil Trans 94:1–16

Zhu Y, Inada H, Nakamura K, Wall J (2009) Atomic imaging with secondary electrons. Nature Mater 8:808–812

Other Documents of Interest Available on the Web at No Charge

Ward et al (2006) Systems and methods for a gas field Ion microscope. US Patent 7,601,953

Ward et al (2008) Atomic level ion source and method of manufacture and operation. US Patent 7,368,727 B2

Ward et al (2012) Ion sources, systems and methods. US Patent 8,110,814 B2

Ward et al (2009) Systems and methods for GFIS. Patent Application US2008/0217555A1

Index

A

Aberrations, 1, 14, 27, 36
Accelerator module, 12
Alignment, 12, 13
ALIS, 7, 9, 11
Aperture, 10, 12, 13, 27
Argon, 3, 51

B

Backscattered ions, 22, 43
Backscattering, 18, 24, 25
Beam range, 15, 18, 33
Best imaging voltage (BIV), 11
Bethe, 21
Brightness, 5, 11, 14, 21, 30, 31, 36, 52

C

Charging, 3, 39–41
Chromatic aberration, 11
Chromatic energy spread, 11
Cold field electron gun (CFEG), 5

D

Damage, 3, 6, 30, 35, 39, 41, 49, 51, 52
Detector, 27, 30–33, 35, 36, 44, 46
Detector quantum efficiency (DQE), 35
Diffraction, 13, 27
Dislocations, 30
Dual Beams, 7, 46

E

Electrons, 1, 2, 15, 18, 20–25, 28, 30–33,
    39–41, 43, 49
Everhart-Thornley (ET) detector, 30, 35
Extractor module, 12

F

Field ion microscope (FIM), 6, 9

G

Ga+, 1, 3, 7, 23, 24, 46–48, 53
Gallium (GFIS), 1, 3, 6, 7, 11, 36, 37, 47, 51,
    52
Gas injection, 41
Graphene, 33, 49, 53

H

Helium, 1, 6, 7, 9, 10, 14, 20, 22, 27, 39, 43,
    49, 55
HIM, 9, 11, 12, 14, 27, 30, 35–37, 43, 46, 47
Hydrocarbon, 6

I

Interactions, 17, 20, 34
Interaction volume, 24
Ion channeling, 28, 30
Ion-generated secondary electrons (iSE),
    18–20, 22, 23, 25–28, 30, 35, 39, 51

IONiSE, 22, 33, 60
isotopes, 52

K
Kanaya-Okayama (K-O), 17, 18

L
LaB6 source, 11
Levi-Setti, 6

M
Mean free path, 22
Microanalysis, 8, 14, 15, 31, 36, 43, 45, 47, 48
Microscopy, 3, 5, 8, 39, 46, 55
Monte Carlo modeling, 22
Muller, 6, 9

N
Nanofabrication, 55, 57
Neon, 1, 47, 51, 52

O
ORION, 7, 12, 36, 47, 52

P
Patterning, 49, 55
Polymers, 1, 45, 49
Protons, 3, 6, 51

R
Resolution, 1, 3, 6, 8, 11, 13, 14, 19, 20,
   23–25, 33, 36, 43, 44, 47, 51, 52, 55
Rutherford backscattering
   (RBS), 30, 31, 44, 45

S
Scanning electron microscope (SEM), 1, 18,
   21, 27, 34, 37, 39, 43, 46
Scanning field ion microscope (SFIM), 9, 10
Secondary electrons (SE), 3, 14, 18, 20–25, 30,
   33, 40
Secondary electron yield, 3, 51
Secondary ion mass spectrometry (SIMS), 45,
   46, 48, 57
Semiconductors, 1, 7, 27
SiDD (silicon drift detector), 31
Spherical aberration, 11
SRIM simulation, 20
Stopping power, 20–22, 33

T
Thickness fringes, 34
Time of flight (TOF), 45–47, 55
Topography, 27
Transmission electron microscope (TEM), 14,
   34, 55
Transmission ion microscopy, 32
Trimer, 10, 11

V
Vacuum, 5, 10, 22

W
Wavelength, 1, 5, 13, 14, 30, 34
Working distance, 36

Z
Zeiss, 7, 12